Preface

The 'Fuzzy Logic' research group of the Microelectronics Institute of Seville is composed of researchers who have been doing research on fuzzy logic since the beginning of the 1990s. Mainly, this research has been focused on the microelectronic design of fuzzy logic-based systems using implementation techniques which range from ASICs to FPGAs and DSPs. Another active line was the development of a CAD environment, named Xfuzzy, to ease such design. Several versions of Xfuzzy have been and are being currently developed by the group. The addressed applications had basically belonged to the control field domain. In this sense, several problems without a linear control solution had been studied thoroughly. Some examples are the navigation control of an autonomous mobile robot and the level control of a dosage system.

The research group tackles a new activity with the work developed in this book: the application of fuzzy logic to video and image processing. We addressed our interest to problems related to pixel interpolation, with the aim of adapting such interpolation to the local features of the images. Our hypothesis was that measures and decisions to solve image interpolation, which traditionally had been done in a crisp way, could better be done in a fuzzy way. Validation of this general hypothesis has been done specifically in the interpolation problem of video de-interlacing. De-interlacing is one of the main tasks in video processing. It is necessary whenever the transmission standard uses an interlaced format but the receiver requires a progressive scanning, as happens to LCDs and plasma displays, projectors, and DVDs. Besides, de-interlacing is usually the first step before applying conversion between two formats, a practice in crescent use due to the proliferation of different video formats.

With the main goal of developing fuzzy logic-based algorithms capable of improving state-of-the-art algorithms for video de-interlacing, the following sub-goals have also been pursued:

- To obtain a simple hierarchical solution which gains efficiency from the combination of its simple constituents (interpolators), being each interpolator specialized in one of the three key image features for de-interlacing: motion, edges,

and possible repetition of areas of fields. Thus, a 'divide-and-conquer' strategy is followed.
- To take inspiration for each interpolator from well-known crisp algorithms so as to make apparent the advantages of using fuzzy processing.
- To apply heuristic knowledge expressed linguistically as the starting point to design the rules of the interpolators, thus exploiting the ability of fuzzy logic to cope with symbolic knowledge.
- To apply a tuning process of the parameters of the rules so as to minimize the error between progressive (ideal) results and de-interlaced results, thus taking advantage of the other side of fuzzy rule bases: its numeric nature.
- To test the performance of the proposed interpolators as well as the final solution with a wide variety of sequences of different formats and origins (video, film, and hybrid).
- To compare our proposal with state-of-the-art algorithms of less, similar, and higher complexity. Not only quantitative evaluations (MSE and PSNR) but also qualitative appreciations will be considered since de-interlaced results will be seen at last by a human.
- To use a computer-aided-design methodology that combines Matlab and its Image Processing to develop image processing algorithms, and Xfuzzy 3 and its tool *xfsl* to tune the parameters of the different fuzzy rule bases proposed.
- To always consider hardware simplicity as a relevant weight when deciding among global (architectural-related) as well as particular (parameter-related) design aspects. These decisions will ease the hardware implementations of the algorithms.

The book is organized in five chapters. In Chapter 1, some basic concepts are explained to completely understand the contribution of the algorithms developed in this research work. The evaluation of how motion is present and how it influences on de-interlacing is studied in Chapter 2. The design options of the proposed fuzzy motion-adaptive de-interlacing algorithm is studied in Chapter 3. A spatial interpolator that adapts the interpolation to the presence of edges in a fuzzy way is developed in Chapter 4. A temporal interpolator that adapts the strategy of the interpolation to possible repetition of areas of fields is presented in Chapter 5. Using both interpolators in the fuzzy motion-adaptive algorithm described in Chapter 3 clearly improves the de-interlaced results.

Seville, Spain
September 2009

Piedad Brox
Iluminada Baturone
Santiago Sánchez-Solano

Acknowledgements

This book is the result of the work developed by a group of researchers that belong to the University of Seville and the Microelectronics Institute of Seville (IMSE-CNM, *http://www.imse-cnm.csic.es/*), which is part of Centro Nacional de Microelectrónica (CNM). The authors would like to express their acknowledgments to the following research projects: the Spanish MEC Project (TEC2008-04920), and the Project (P08-TIC-03674) from the Andalusian regional Government. P. Brox was financially supported by the FPU program from the Spanish ministry of Science and Education for young postgraduate students. Finally, the authors would also like to thank other colleagues of 'Xfuzzy team' since they have helped and encouraged the research work developed in this book.

Contents

Chapter 1
Basic Concepts

Abstract. In this Chapter, some basic concepts are presented to completely understand the contribution of the algorithms developed in this research work. The Chapter is organized in six sections. In Section 1.1 the television transmission systems are introduced from a historical perspective. Video material comes from different kinds of sources that are described in Section 1.2. De-interlacing is demanded by modern consumer electronic products and the state of the art of video de-interlacing algorithms is presented in Section 1.3. The alternatives to develop the hardware implementation of de-interlacing algorithms are described in Section 1.4. In Section 1.5, the role of fuzzy logic in video processing during the last years is analyzed. Finally, the conclusions of this Chapter are expounded in Section 1.6.

1.1 Television Transmission Systems

1.1.1 Monochrome Television Systems

The first steps of the television commercialization took place in the late 1930s. The Radio Manufacturers Association (RMA) in the United States of America (USA) set up a committee that recommended a primitive standard in 1937. However, there was no a general agreement until the creation of the National Television System Committee (NTSC) in 1940. The final report of the NTSC was officially adopted by the Federal Communications Commission (FCC) in 1941 and fixed the NTSC monochrome standard. Its main characteristics can be summarized as follows [1]:

- The use of a 6 MHz radio frequency (RF) channel with the picture carrier 1.25 MHz above the bottom of the channel and the sound carrier 4.5 MHz above the picture carrier.
- Each frame consists of 486 visible scan lines of a total of 525 lines.
- 4:3 aspect ratio. It is the ratio of the longer dimension of a frame to its shorter dimension.
- Interlaced scanning. From the beginning of public television service, a technique of interlaced scanning was employed. It divides the scanning pattern into two sets

P. Brox et al.: Fuzzy Logic-Based Algorithms for Video De-Interlacing, STUDFUZZ 246, pp. 1–48.
springerlink.com

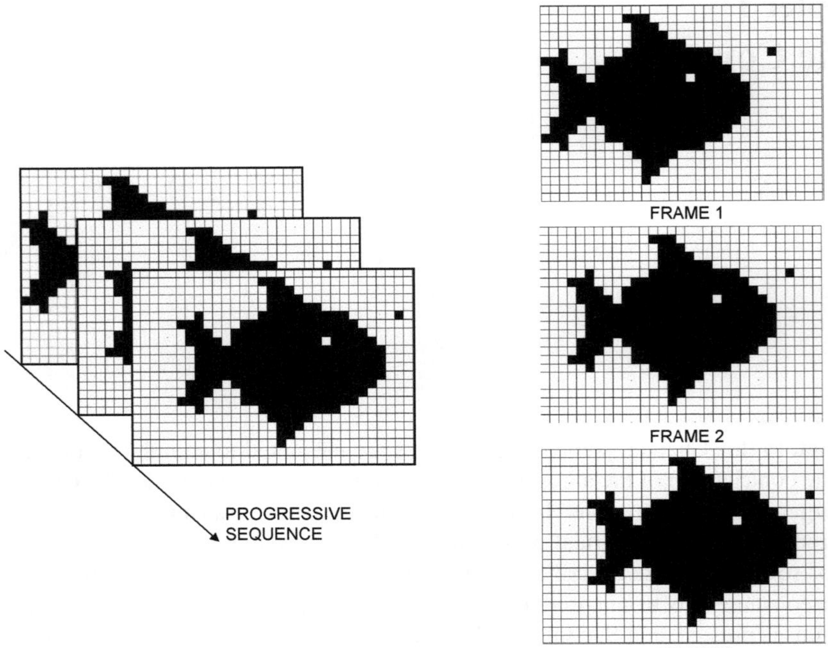

Fig. 1.1. Progressive video sequence of a moving fish.

('odd' and 'even') of spaced lines that are displayed sequentially. The transmitted image only contains the odd or even lines and is called field instead of frame. The interlaced scanning or interlacing was introduced to reduce video bandwidth without degrading in excess the picture quality. The afterglow of the phosphors in the Cathode Ray Tubes (CRTs) makes that these devices can display interlaced video directly. Therefore, interlacing is a smart mechanism to provide a full horizontal detail with the half of the bandwidth.

- A refresh frequency of 30 frames (60 fields) per second. This frequency was selected because it matched the nominal 60 Hz frequency of alternating power used in the USA. This avoids wave interference that produces rolling bars on the screen display. Furthermore, it is enough to prevent visible flicker.
- 15.75 kHz line frequency. It is the result of multiplying 60 fields per second by 262.5 lines per field.

Figure 1.1 shows a fictitious progressive sequence of a moving fish. The corresponding interlaced sequence is shown in Figure 1.2. Fields are classified in the order they arrive, and are categorized by an odd or even polarity. As shown in Figure 1.2, even fields only contain the even-numbered lines whereas odd fields contain the odd-numbered lines.

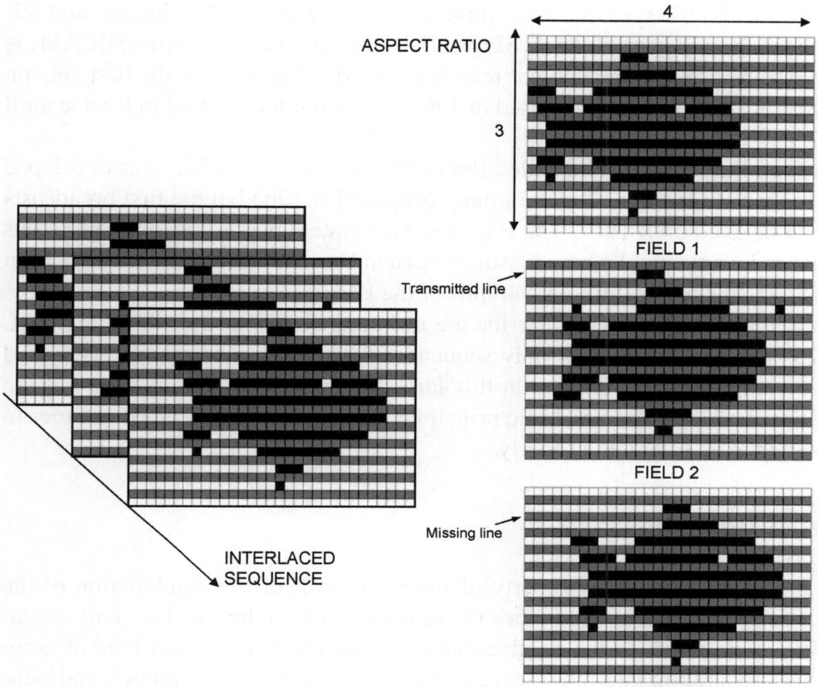

Fig. 1.2. Interlacing the sequence in Figure 1.1.

1.1.2 Analog Color Television Broadcast Systems

Color was demanded by the public since the beginning of the development of commercial television systems. Nevertheless, the engineering community required almost ten years to tackle this problem. To achieve a successful color television system, it was essential that the color system was fully compatible with the existing monochrome system. This meant that existing terminal equipment and receivers must be able to accept color transmissions.

Most of the colors found in nature can be approximated by a color reproduction system developed and codified during the 1920s and 1930s by the Committee International d'Eclairage (CIE). The CIE basis was studied in the 1950s by NTSC during the development of color TV in the USA. Finally, FCC approved transmission standards and authorized broadcasters in the USA under the CIE principles in 1954 [1]. Subsequently, this system was also adopted by Canada, Japan, Mexico, and other countries.

European countries delayed during some years the adoption of a color television system. The already existing NTSC standard was not compatible with the 50 Hz frequency of the grids in Europe. The development of this new standard had to provide a picture rate of 50 fields per second but also, it could be interesting to improve some weaknesses of NTSC system, for example, the use of wide band luminance and relatively narrow band chrominance. SEquential Couleur Avec Memorie (SECAM) is historically the first European color television standard. In spite of the first version of the SECAM system was proposed in 1961, it was not transmitted in France until 1967.

A second European system called Phase Alternating Line (PAL) was developed at Telefunken in Germany. It was initially proposed in 1963 but the first broadcasts were also delayed until 1967. The PAL standard system was adopted by numerous countries in Europe. Public broadcasting began in Germany and United Kingdom in 1967 using two slightly different variants of the PAL system.

The following subsections describe the technical details of the NTSC and PAL color TV standards. Although only sequences from PAL system have been used to test the approaches developed in this PhD. dissertation, NTSC system was the first standard and stated some basic principles that have to be described in order to understand the PAL system properly.

1.1.2.1 NTSC Color System

Color is a psycho-physical property of light, as result of the combination of the characteristics of light that produces the sensations of brightness, hue, and saturation. Brightness refers to the relative intensity; hue refers to the attribute of color that allows separation into three spectral groups (red, green, and yellow); and saturation is defined as the degree of a color deviates from a neutral gray of the same brightness and, according to it, the color is classified as pure, pastel or vivid. These three characteristics cover the total information to define a color. This description is based on the CIE principles and was used by engineers to develop encoding and decoding color techniques.

NTSC uses a luminance-chrominance encoding system. Two signals are transmitted: luminance which transports the brightness and occupies the wide-band portion of the previous monochrome channel, and chrominance which represents the hue and saturation attributes of light. Thus, chrominance carries color information while the system is fully compatible with the existing black and white system.

The color encoding technique used by the NTSC system was different to the system used in video cameras that translate incoming light into electrical signals that correspond to the three primary colors Red, Green and Blue (RGB) parts. Conversion from RGB electrical signals to luminance-chrominance encoding system is straightforward without loss of information except for possible round-off errors [2].

The three systems NTSC, PAL, and SECAM use this luminance-chrominance encoding system but they differ from the transmission and modulation of the chrominance signal.

The NTSC color standard occupies a total bandwidth of 6 MHz. The video carrier is transmitted 1.25 MHz above the lower bound of the channel. Engineers chose a frequency close to 3.579 MHz for the chrominance subcarrier in order to minimize the interference between the luminance and the chrominance signals [1]. The main frequency of the audio carrier is equal to the original used in the NTSC monochrome standard (4.5 MHz). Thus, the existing receivers were capable of tuning the audio signal properly. Nevertheless, the line frequency was modified up to 15.734 kHz in order to prevent visible beating between the color and sound carries, and this changed the field frequency to 59.94 Hz, that is, the result of dividing the line rate by the number of lines in a field ($\approx$15.734 KHz/262.5).

1.1.2.2 PAL Color System

Except for some minor details, the color encoding principles for PAL are essentially the same as the used in the NTSC system. However, the phase of the color signal is reversed by 180° from line-to-line and this is why is named Phase Alternating Line (PAL). This is done for the purpose of averaging, or canceling, certain errors resulting from amplitude and phase distortion of the color modulation subcarrier. The frequency of the color subcarrier in PAL system is increased up to 4.43 MHz approximately.

1.1.2.3 Summary and Comparisons of NTSC, PAL, and SECAM Standards

The three conventional television broadcast signals transmitted throughout the world have the following similarities :

- All are compatible with the monochrome broadcasting standard.
- All the systems use two fields interlaced to create a complete frame.
- All the standards set an aspect ratio 4:3.
- All employ a luminance and a chrominance signal.
- The sound is transmitted on a higher frequency than the visual carrier.

The main difference among these three standards is that they are currently used in different countries:

- NTSC is used in the USA, Canada, Central America, most of South America, and Japan.
- PAL is used in the United Kingdom, many western European countries, and China. There are several variations of the PAL system.
- SECAM is used in France, countries and possessions influenced by France, Russia and most of the ex-Soviet nations, and other areas influenced by Russia.

Figure 1.3 shows a map of the world in which the standard used in each country is illustrated.

The three standards are incompatible for the following reasons:

- The line structure of the signals varies. NTSC uses 525 lines per frame, 30 frames or almost 60 fields per second (59.94 Hz). PAL and SECAM use 625 lines per frame, 25 frames or 50 fields per second.
- The line or horizontal frequency also varies. For the NTSC system, it is 15.734 kHz. For the PAL and SECAM system, it is 16.625 kHz.

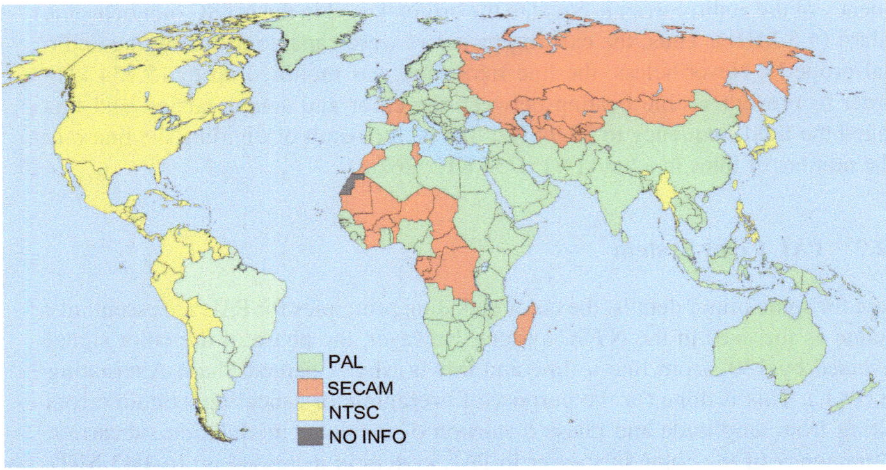

Fig. 1.3. Standard for color television broadcast used in each country.

- Channel bandwidths are different. NTSC uses a 6 MHz channel width. Several versions of PAL exist with 6, 7, and 8 MHz bandwidths. SECAM channel is 8 MHz wide.
- Video signal bandwidths are different. NTSC uses 4.2 MHz. Different PAL versions use up to three different frequencies 4.2, 5, and 5.5 MHz. SECAM uses 6 MHz.
- The color subcarrier signals are incompatible. NTSC uses 3.579 MHz. PAL uses 4.433 MHz. SECAM utilizes two subcarries, 4.406 and 4.25 MHz.
- The offset between visual and audio carries varies. In NTSC, it is 4.5 MHz. In PAL, the separation is 5.5 or 6 MHz, depending upon the PAL type. SECAM uses 6.5 MHz.

It is possible to convert from one television standard to another electronically. The most difficult part of the conversion process results from the different number of scan lines. Complex computer algorithms compare information on pairs of lines to determine how to create the new lines required (for conversion to PAL and SECAM) or how to remove lines (for conversion to NTSC). Static elements in the picture present no great difficulties, but motion in the picture may produce objectionable artifacts as the result of the sampling system.

Let us illustrate the differences among the three systems to encode color by showing the CIE chromaticity diagram. This is a two-dimensional plot to define color as shown in Figure 1.4. The parameters x and y are defined as the two color coordinates, which specify one point in the chromaticity diagram. The third coordinate is the luminance, which corresponds to the brightness attribute. Numbers in the plot give the wavelength of light in nanometers.

Figure 1.4 shows three inscribed triangles. The triangle called CRTs represents the red, green, and blue locations of emission phosphors used in a CRT, and defines

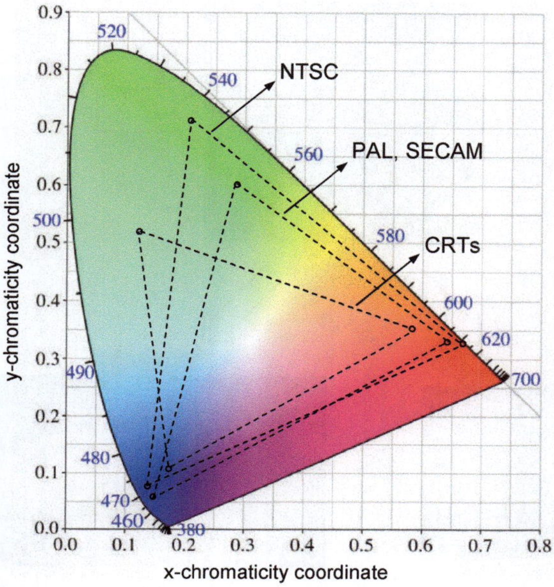

Fig. 1.4. CIE chromaticity diagram comparison of systems.

all the colors that the tube can display. Some colors can not be generated by mixing the three phosphor colors, and thus, they are drawn outside the triangle. Figure 1.4 also shows the primary coordinates for the NTSC, PAL, and SECAM standards. As shown in CIE chromaticity diagram, the three standards cover different colors.

There are a number of organizations responsible for standardization at global, national, and regional levels. In the broadcast television field, one of the most prominent is the International Radio Consultative Committee (CCIR). CCIR was a branch of the International Telecommunication Union (ITU), which is part of the United Nations Organization. In 1993, the CCIR was renamed as ITU-R (the Radiocommunication Sector of the ITU). The crucial documents issued by ITU-R are recommendations and reports that contain explicit information for different variants in practical implementation [3]. More details on these standards are described in [1], [4]-[5].

1.1.3 Digital Color Television Broadcast Systems

The decades after World War 2 saw a rapid and widespread development of computing technology mostly for military purposes. The first digital computer systems using binary code soon became dominant. The introduction of integrated semiconductor logic circuits in the late 1950s and the reduction of the manufacturing costs encouraged the commercial applications of digital systems. In the mid-1960s broadcasters also began to experiment with digital coding of sound and then, later, TV signals began to exploit the advantages of the new format.

Fig. 1.5. Differences in the scene capture capabilities of conventional video and HDTV.

Analog signal coding and transmission are prone to a number of defects. By embodying the information to be transmitted or stored in waveforms any distortion of the waveform in transit is a distortion of the information. Digital representation does not suffer these defects and information transmission and storage becomes better by using binary arithmetic computing techniques. On one hand, one disadvantageous consequence of using digital systems is that the rate at which bits have to flow is quite high, typically hundreds of Mega-Bits per second, and thus needs a high bandwidth to convey it[1]. On the other hand, the robustness of the digital format is its main advantage. Any damage to the bit stream during transmission, for instance bit errors, can be managed by means of error detection and correction codes. Furthermore, analog systems normally use time-dependent parameters that require frequent maintenance.

The success of digital transmissions was not only its ability to avoid analog defects but engineers also wanted to enlarge the visual field occupied by the video image. Traditional or analog Standard Definition TV (SDTV) formats (NTSC, SECAM and PAL) offer a limited scene content as illustrated in Figure 1.5. The angle of view in a SDTV system is narrow and forces the camera operator to present a 'small window' of the scene as shown in Figure 1.5. Let us compare the viewing angle for a SDTV

[1] This problematic encouraged the development of video compression techniques to reduce the bandwidth required to transmit digital video [6].

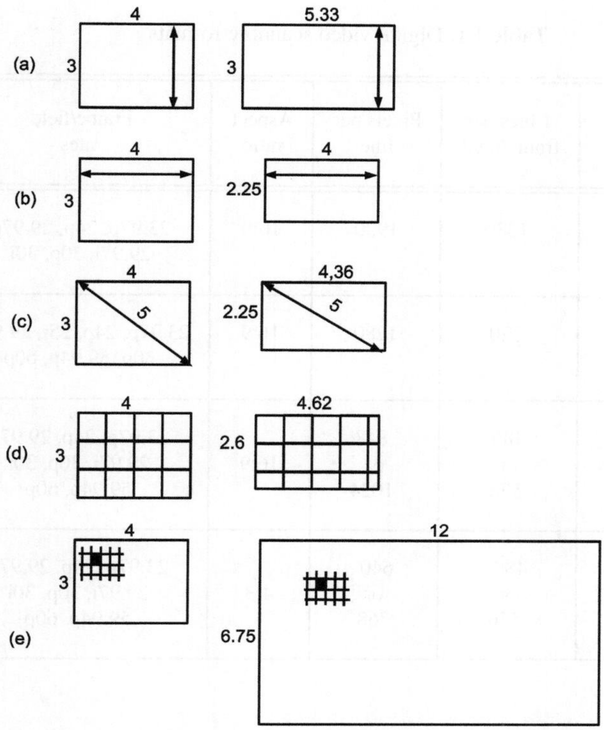

Fig. 1.6. Comparison of aspect ratios between SDTV and HDTV.

system with the corresponding to a HDTV system. As indicated in Figure 1.5, a direct consequence is the increase of image content of the captured scene.

A novelty introduced by the HDTV systems is the increase of the aspect ratio from 4:3 to 16:9. Comparisons between a SDTV and a HDTV system were made in Figure 1.6 by using various measures: equal height (a), equal width (b), equal diagonal (c), and equal area (d). All these comparisons are wrong since they overlook the fundamental improvement of HDTV, the increasement of pixel count. The correct comparison is performed in Figure 1.6(e). This illustration shows the enlargement of the screen picture in a HDTV system when both frames contain the same picture detail.

The first developmental work on a HDTV system began in 1968 at the technical research laboratories of Nippon Hoso Kyokai (NHK) in Tokyo [7]. Nevertheless, the CCIR did not begin to study HDTV up to the mid-1980s, and defined HDTV systems as those with 1000 pixels per line or more.

Although early HDTV formats in Europe and Japan were analog, HDTV is usually broadcast digitally since video compression techniques provide reasonable bandwidths. Currently the majority of HDTV standards are based on the use of the

Table 1.1. Digital video scanning formats

Definition	Lines per frame/field	Pixels per line	Aspect ratio	Frame/field rates
High (HD)	1080	1920	16:9	23.97p, 24p, 29.97p 29.97i, 30p, 30i
High (HD)	720	1280	16:9	23.97p, 24p, 25i, 29.97p 30p, 59.94p, 60p
Standard (SD)	480 or 576	848 or 1024	16:9	23.97p, 24p, 29.97p 29.97i, 30p, 30i 59.94p, 60p
Standard (SD)	480 or 576	640 or 768	4:3	23.97p, 24p, 29.97p 29.97i, 30p, 30i 59.94p, 60p

Fig. 1.7. Comparison SDTV vs. HDTV using a snapshot of a real sequence from NBC channel.

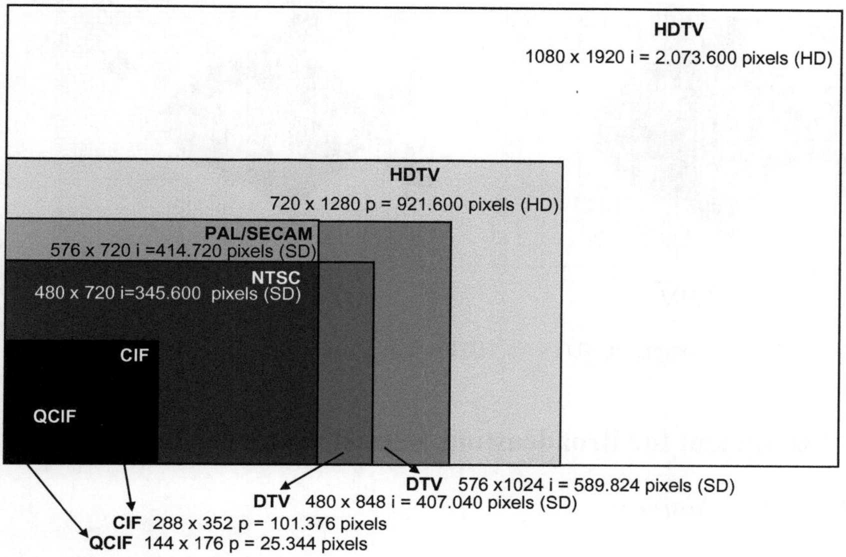

Fig. 1.8. Comparison of TV formats.

MPEG-2 compression algorithm, 2nd generation source coding algorithm developed by the Moving Pictures Expert Group [6].

Several HDTV standards are currently defined by the ITU(ITU-R BT.709 recommendation) [3] as shown in Table 1.1. They include 1080i (1080 actively interlaced lines), and 720p (720 progressively scanned lines). All of HDTV standards use a 16:9 aspect ratio. As illustrated in Figure 1.7, the change of aspect ratio from 4:3 to 16:9 provides an increase of the picture area. Figure 1.7 corresponds to a real snapshot from a TV sequence of the NBC channel.

Currently, broadcast industry works with the DTV formats shown in the list of Table 1.1 [7]-[8]. Each standard is characterized by a frame rate if the material is progressive (p), or a field rate if the material is interlaced (i).

Figure 1.8 establishes a comparison among resolutions of several TV formats: the three analog video standards (NTSC, SECAM and PAL), DTV and HDTV standard systems. Common Intermediate Format (CIF) shown in Figure 1.8 is a video-coding standard developed to support videoconferencing schemes. CIF format complies with the ITU H.621 video-conferencing standard [3]. It was designed to be easily converted to PAL and NTSC standards. It specifies a data rate of 29.97 frames/s, which each frame containing 288 lines and 352 pixels per line. A related standard is the Quarter CIF (QCIF), which implies the height and width of the CIF frame are halved.

HDTV has at least twice the linear resolution of SDTV, thus offering much more details. This is shown in Figure 1.9 where the resolution of a fictitious image is doubled.

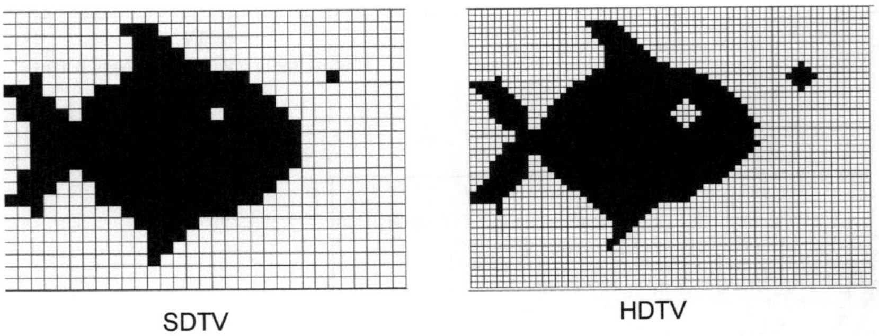

Fig. 1.9. Comparing SDTV vs. HDTV format using a fictitious image.

1.2 Equipment for Broadcasting Television Images

1.2.1 Video Cameras

Currently there are two types of professional video cameras: high-end portable and studio cameras. Portable professional cameras are larger than consumer cameras (camcorders) and are designed to be carried on the cameraman's shoulder. On the contrary, studio cameras are on studio, usually with pneumatic or hydraulic mechanisms to adjust the height and some of them are on wheels to facilitate its mobility.

A video camera uses a sensitive electronic tube to transform light into electrical impulses that are transmitted to TV receivers. Early TV cameras used tubes, called iconoscope, required critical conditions of light. This limitation was overcome with the development of tubes. Figure 1.10(a) corresponds to a TV camera from 1956. Tubes were substituted for Charge-Couple Device (CCD) in the 1980s. Nowadays the most advanced cameras use three CCDs to achieve a better color separation. Three-CCD or 3CCD cameras have three separate CCDs, each one taking a separate measurement of red, green, and blue light. Light coming into the lens is split

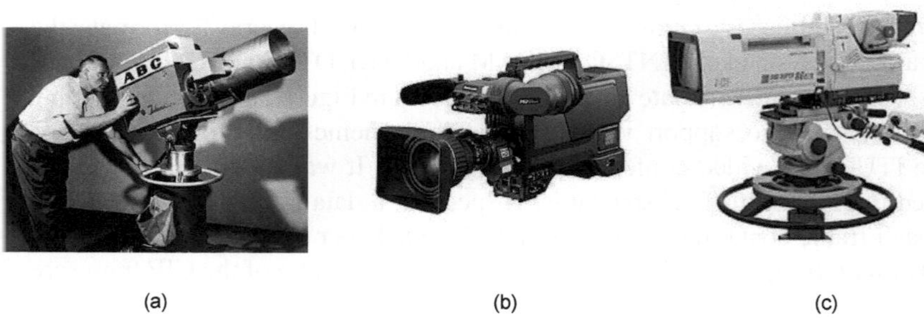

Fig. 1.10. (a) An early professional video camera from 1956. (b) A portable professional camera. (c) A current studio professional camera.

(a) (b)

Fig. 1.11. (a) Bolex Paillard H8 Reflex film camera. (b) Arrican Lite, a popular 35 mm film camera.

by a trichroic prism assembly, which directs the appropriate wavelength ranges of light to their respective CCDs. Three-CCD cameras are generally more expensive than single-CCD cameras because they require three times more elements to form the image detector, and because they require a precision color-separation beam-splitter optical assembly. Other type of cameras is the Complementary Metal Oxide Semiconductors (CMOS) cameras that use image sensors of CMOS technology. Although they are less popular than CCDs in high-performance professional and industrial cameras, CMOS sensors are more demanded in low-cost low-power consumer applications (camcorders, cellphones or web cameras) since their integration capability is higher. Furthermore, additional circuitry on the CMOS image sensor may be included to perform a local processing the captured images.

Figure 1.10(b) shows a portable HDTV camera. It is a AK-HC931B from Panasonic that delivers HD images in 1080i and 720p formats. Figure 1.10(c) shows a studio camera. Exactly it is a Sony HDC-1000 that utilizes advanced HD Digital Signal Processing with 14 bit A/D. The Sony HDC-1000 creates images in either 1080p /50 Hz frame progressive or 1080i /50 Hz field interlace frame rates, depending on the desired final system output.

1.2.2 Movie Cameras

Movie or film cameras are a type of photographic camera used in film industry to take a rapid sequence of photographs on strips of film. They have exposure control via an iris aperture located on the lens. Also, there is a rotating system to pass the light from the lens to the film, or reflects it into the viewfinder. Figure 1.11(a) shows the H8 movie camera from Bolex Paillard invented in 1936. It was equipped with reflex viewfinder and it can receive a double-8 film, and thus it gives out 60 meters film (16 minutes approximately). Figure 1.11(b) shows a current movie camera from Arri called 'Arricam Lite' that is a complete 35 mm film production system with modern features. Although the 16 mm width was the most common format for distribution for many years, 35 mm is currently the most used format. The standard frame rate for film production systems is 24 frames per second.

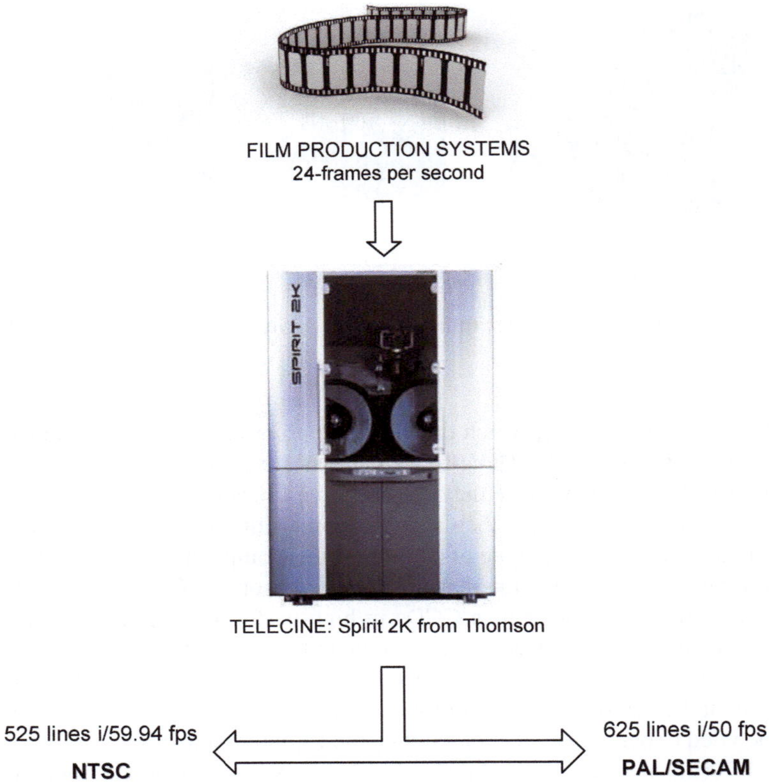

Fig. 1.12. Diagram flow of the telecine conversion process.

1.2.3 Film-to-Video Transference: Pull-Down Process

Film is a program production standard with no direct compatible relationship with any TV standard. Therefore the process of standard conversions is a fundamental process. A 24-frame celluloid medium must be converted to a 25-frame/50-field (PAL) or 29.97-frame/59.94-field (NTSC) TV medium, depending on the region of the globe, for conventional television broadcast. This conversion is performed by means of well-known standards converters or telecine machines.

Figure 1.12 illustrates the process using the most popular telecine machine called Thomson Spirit DataCine. This telecine works with 35 mm film material but can also scan 16 mm, and enables a motion picture to be compatible with all television standards (PAL, NTSC, SECAM, HDTV).

The most complex part of telecine is the synchronization of the mechanical film motion and the electronic video signal. Every time the video part of the telecine samples the light electronically, the film part of the telecine must have a frame ready to be photographed. This would be easy when the film is photographed at the same frame rate as the video camera will sample, but this is not the reality and a procedure is required to change frame rate. The process of converting film material to conventional TV standards is called pull-down process.

Pull-down is a simple process where the previous image from the film is repeated until a new one is available. This method can easily be implemented mechanically, being the preferred method to transfer film to video. The general public accepts the motion artifacts introduced by this method since these motion artifacts are also presented in the cinema, where the image is repeated multiple times to reduce large area flicker. The art director also minimizes these artifacts by employing techniques to reduce them such as using tracking shots (keeping the same position of the camera in respect to the foreground object), long exposure times (the artifacts are reduced by blurring) and using small focus depth (the static foreground objects are in focus, but the moving background is out of focus, hiding the artifacts).

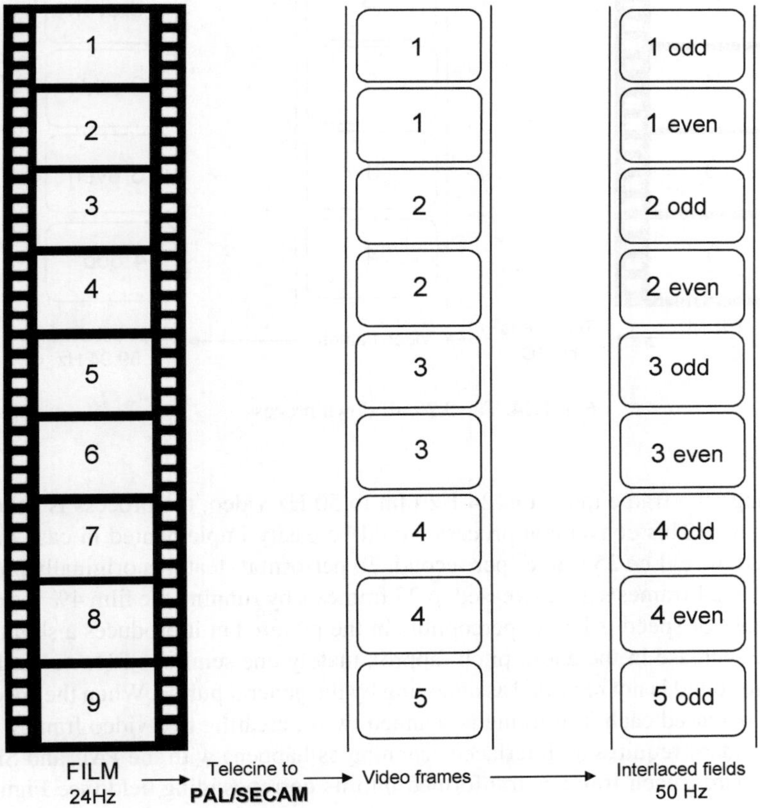

Fig. 1.13. The 2:2 pull-down process.

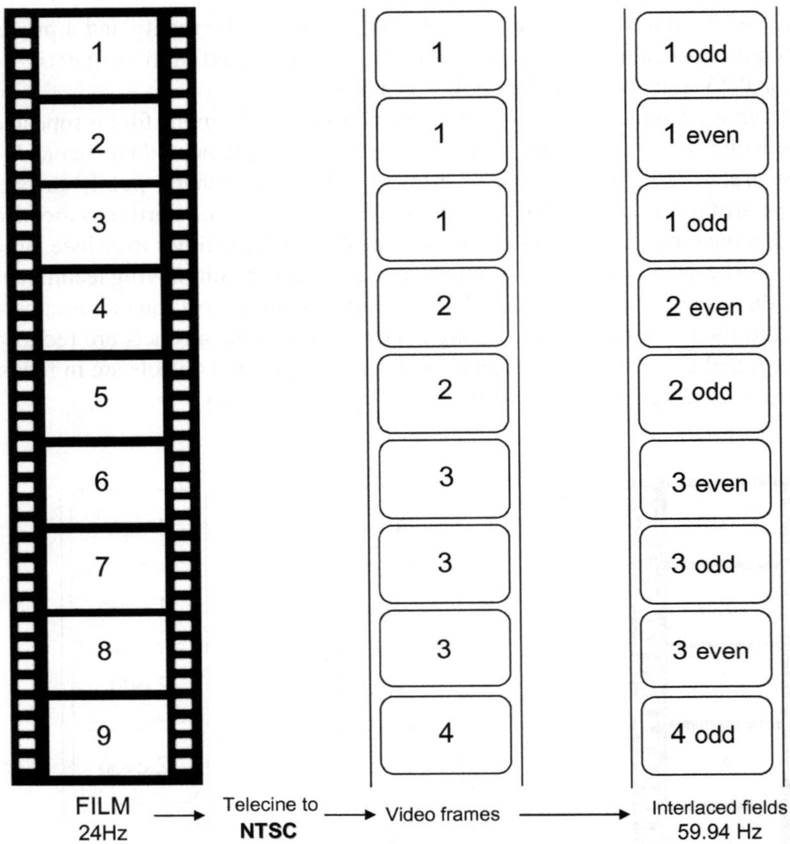

Fig. 1.14. The 3:2 pull-down process.

To change the frame rate from 24 Hz film to 50 Hz video, the process is called 2:2 pull-down. This conversion process would be easily implemented in case that the film rate would be 25 frames per second. To perform it, features originally photographed at 24 frames/s are increased to 25 frames/s by running the film 4% faster. This increase of speed is barely perceptible in the picture but it produces a slightly noticeable increase in the audio pitch (approximately one semitone). However the pitch of the sound is not regarded as annoying by the general public. When the speed has been increased each film frame is scanned twice, creating two video frames. If the TV standard requires an interlaced scanning as happens with the PAL and SECAM standards, each frame is tranformed into its correposnding field (see Figure 1.13).

To transfer 24 Hz film to NTSC format, the increase of speed up to 30 Hz is not desired since it causes a distortion in the picture and the pitch of the sound that is regarded as unacceptable by the general public. Therefore another method is used.

The first step is to slow down the film motion by a factor 1/1.001, and thus the film rate is decreased from 24 Hz up to 23.976 Hz. This speed change is unnoticeable to the viewer. After this, each even frame is repeated three times while every odd film frame is repeated twice as shown in Figure 1.14. This creates an increase of frame-rate by a factor 2.5, obtaining a video signal with a rate of 59.94 fields per second. This method is called 3:2 pull-down.

1.3 The Need of De-Interlacing

The success of the interlaced video scan format is based on the technologies used to implement TV in the analog area, since CRTs display de-interlaced video directly. Although present-day technologies are enough powerful to achieve progressive video signal, interlacing is currently used by all the analog TV broadcast systems (NTSC, PAL and SECAM) and also by some of modern digital transmissions as was described in Section 1.1.

The weakest points of the interlaced scanning format are the degradation of image quality and the increase of complexity of scanning format conversion.

Interlaced scanning produces annoying artifacts such as line flicker, interline twitter, and line crawling. Since the total area of the image flashes at the rate of the field scan (50 Hz or 59.94 Hz, depending on the standard), the individual lines repeat at the slower frame rate given rise to several degradations associated with the effect known as interline flicker. Interline twitter is a specific kind of flicker that occurs in fine horizontal lines, for instance, horizontal outlines of an object. Interlacing displaces these lines up and down. Line crawling happens when diagonal edges move slowly in the vertical direction. The displayed line seems to crawl across the edge that is being tracked.

Scanning format conversion is considerably more complex if video is interlaced. Conversion between formats was required in the past by international programme exchange. Nowadays, this demand has recently increased due to the advent of new video scanning standards: high-definition television, videophone, internet and video on PCs. Furthermore, interlaced video signal is unsuitable for a large amount of current devices such as DVDs, projectors, plasma, and LCD display, which need a progressive scanning format.

Both facts (the increasing need of the conversion between formats together with the requirements in consumer equipments) have encouraged the development of de-interlacing algorithms. They perform the reverse operation of interlacing with a frame rate equal to the original field rate. At the same time, they try to achieve an improvement of the image quality by minimizing the artifacts introduced by interlacing.

Many de-interlacing algorithms have been proposed for the last two decades. They range from simple spatial interpolation up to advanced motion compensated interpolation algorithms. The following subsection outlines the most relevant de-interlacing techniques presented in recent literature.

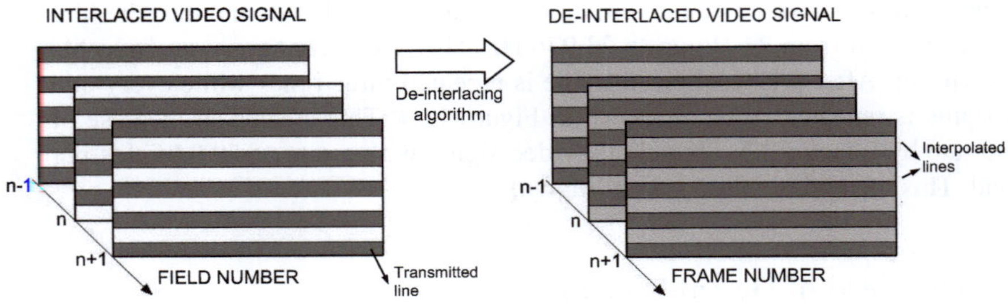

Fig. 1.15. De-interlacing task.

1.3.1 Review of De-Interlacing Algorithms

The task of de-interlacing is shown in Figure 1.15. Fields are classified according to the order in which they are sent. A complete or progressive frame containing odd and even lines is calculated at the receiver using interpolation techniques provided by de-interlacing methods. Figure 1.15 clearly illustrates that de-interlacing process doubles the vertical sampling density. To perform this task, many proposals have been reported in the literature. From simple linear de-interlacing algorithms up to complex motion compensation suitable to tackle moving areas of the picture at expense of a high computational cost.

De-interlacing algorithms can be classified into two categories: non-motion compensated (non-MC) and motion compensated (MC) algorithms [9].

1.3.1.1 Non-Motion Compensated Methods

Non-MC methods are normally simpler than MC methods. They can be divided into linear and non-linear techniques. The first ones always interpolate the same pixels to calculate a new pixel value. On the contrary, non-linear techniques adapt the interpolation strategy to the features in the image. Both categories, linear and non-linear techniques, include spatial (intra-field) , temporal (inter-field), and spatio-temporal techniques [9]. Among spatial methods, the simplest one doubles the line previously transmitted and this is why it is also named line doubling. This method is known as 'Bob' method in the PC community. Figure 1.16 shows how it performs in the fictitious sequence of the moving fish. As it can be seen, it can not reconstruct properly the outline of the fish. This method is usually overcome by line average method that carries out line average of the pixels in the upper and lower lines. The main drawback of line average is that it degrades the sharpness of clear edges as shown in Figure 1.16.

More complex spatial de-interlacing algorithms perform an adaptive interpolation according to the presence of edges. The starting point of this kind of algorithms was presented in [10] and it is called ELA (Edge-based Line Average) algorithm. This initial proposal (also called ELA 3+3) tries to find the local direction of

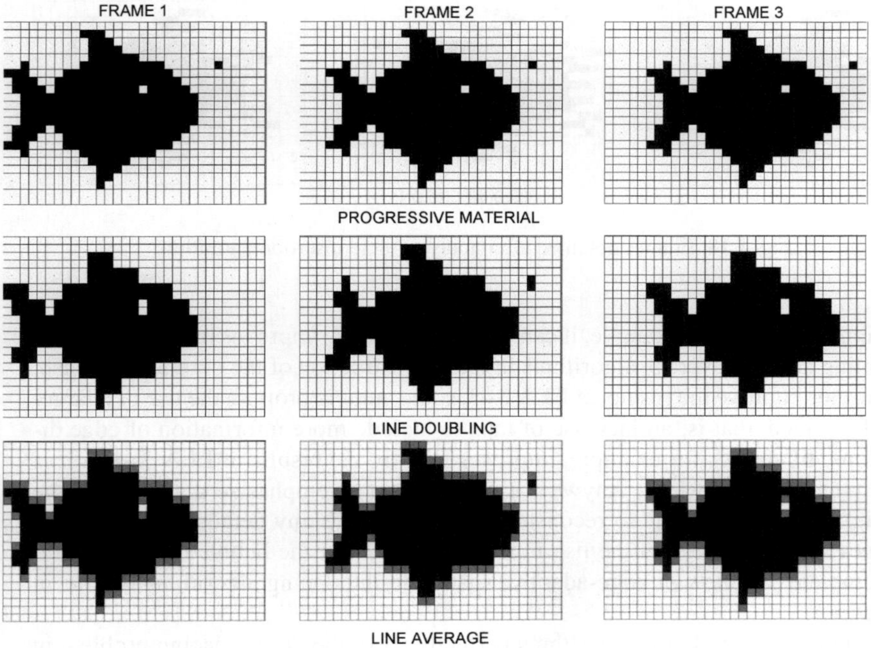

Fig. 1.16. Illustrating line doubling and line average de-interlacing methods.

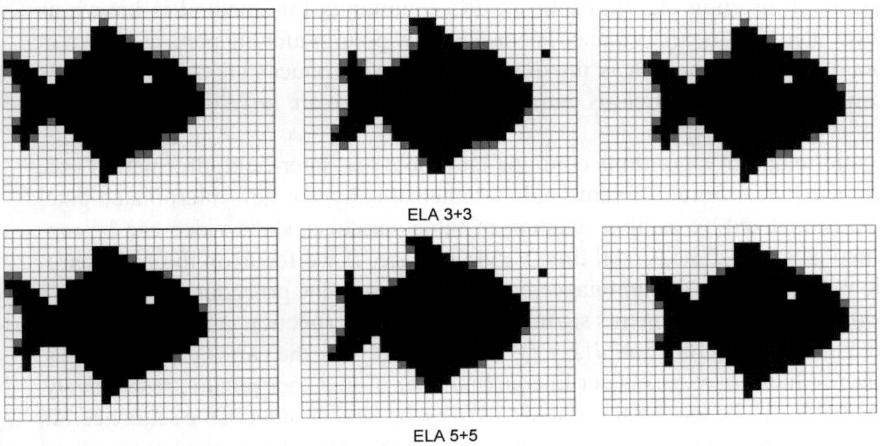

Fig. 1.17. De-interlaced frames obtained after applying ELA 3+3 and ELA 5+5 algorithms.

possible edges using the maximum correlation between three nearest neighbor pixels from the upper line and the other three from the lower line. This algorithm performs well when the edge direction agrees with the maximum correlation as happens

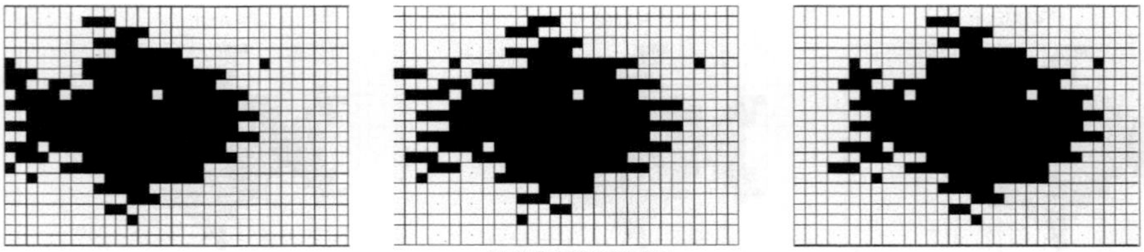

Fig. 1.18. Frames obtained after applying field insertion algorithm.

in Figure 1.17. In this example, it can be seen how ELA improves the results of line doubling and line average algorithms in the reconstruction of the outline of the fish. To further improve the results of ELA 3+3, some authors propose the use of a larger neighborhood, that is, an increase of taps to include more information of edge directions. ELA 5+5, for example, improves slightly the results of ELA 3+3 as it is illustrated in Figure 1.17. Anyway, it is important to emphasize that none of the spatial algorithms is able to reconstruct properly some tiny details of the original images such as the eye of the fish in frames 1 and 3, or the bubble in the frame 2. A complete overview of edge-adaptive spatial de-interlacing algorithms is done in Chapter 4.

Temporal interpolation algorithms perfectly solve the de-interlacing problem in stationary areas of the image. However they produce very annoying effects in moving areas. The simplest temporal de-interlacing method consists of repeating the line from the previous field of the sequence at the current spatial position, it is named field insertion algorithm. In the PC community this method is known as 'Weave' method . Its performance is considerably poor in moving parts as shown in Figure 1.18. Since the fish is moving, this method produces an effect known as 'ghosting' effect. This effect is emphasized when there are large displacements of the objects in the images.

Spatio-temporal algorithms combine intra-field and inter-field information and, hence, use pixels from different fields. Mathematically, the new interpolated pixel can be formalized by using a function of three variables, since video can be represented as a two-dimensional time-varying image. If we focus on the monochromatic video signal or the luminance component signal (the procedure can be easily extended to color components signals), the luminance function can be denoted by a three dimensional function, $I(x,y,t)$, where x and y are the cartesian coordinates, that is, the number of pixels per line and number of lines per picture, respectively. The variable t is the temporal variable and fixes the field number in the sequence. As images are digitized in pixels, x, y, and t are discrete variables defined over natural numbers.

Some spatio-temporal methods employ linear techniques to combine the pixels of the spatio-temporal neighborhood. An example is the vertico-temporal (VT) method reported in [11] that employs the luminance values of pixels in the previous (t-1)

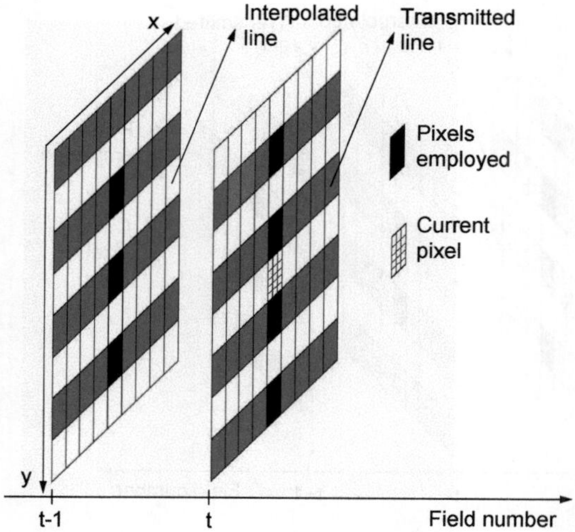

Fig. 1.19. Pixels employed by the VT 2-fields method in [11].

and the current (t) field, as shown in Figure 1.19. This method interpolates the new luminance value by applying the following linear expression:

$$I_p(x,y,t) = \left\{ \begin{array}{l} -\frac{5}{18}I(x,y-2,t-1) + \frac{10}{18}I(x,y,t-1) \\ -\frac{5}{18}I(x,y+2,t-1) + \frac{1}{18}I(x,y-3,t) \\ +\frac{8}{18}I(x,y-1,t) + \frac{8}{18}I(x,y+1,t) + \frac{1}{18}I(x,y+3,t) \end{array} \right\} \quad (1.1)$$

Another example is the VT method reported in [12] that employs the luminance values of pixels in three fields (t-1, t, t+1) according to the following expression:

$$I_p(x,y,t) = \left\{ \begin{array}{l} -\frac{1}{16}I(x,y-2,t-1) + \frac{2}{16}I(x,y,t-1) \\ -\frac{1}{16}I(x,y+2,t-1) + \frac{8}{16}I(x,y-1,t) \\ +\frac{8}{16}I(x,y+1,t) - \frac{1}{16}I(x,y-2,t+1) \\ +\frac{2}{16}I(x,y,t+1) - \frac{1}{16}I(x,y+2,t+1) \end{array} \right\} \quad (1.2)$$

Figure 1.20 shows the larger temporal window used by this linear spatio-temporal method.

The limitation of this kind of VT methods is that they can not detect objects with vertical details that move horizontally since they only consider vertical neighbors.

Diverse non-linear spatio-temporal techniques have been presented in the literature, among which the motion-adaptive approaches have been one of the most relevant proposals. They are based on the fact that linear temporal interpolators are perfect in the absence of motion, whereas linear spatial methods offer a most adequate solution in case that motion is detected. Motion detection can be implicit, as in median-based techniques [13]-[14], or explicit, using a motion detector [15]-[18].

Among motion-adaptive de-interlacing algorithms, the implicit median-based technique that uses a three-point VT median filter has widely been employed, al-

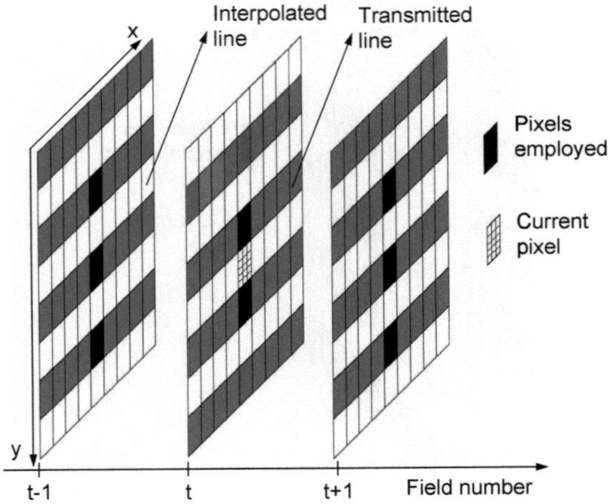

Fig. 1.20. Pixels employed by the VT 3-fields method in [12].

though it has the same limitation of VT methods commented above. The interpolated value is found as the median luminance value of the vertical neighbors in the upper and lower lines, and the temporal neighbor at the same spatial position in the previous field (see Figure 1.21):

$$I_p(x,y,t) = med(I(x,y-1,t), I(x,y+1,t), I(x,y,t-1)) \tag{1.3}$$

The median operator ranks the set of values and returns the value in the middle.

The performance of the explicit methods relies on the quality of the motion detector. Several problems can complicate the motion detection such as the presence

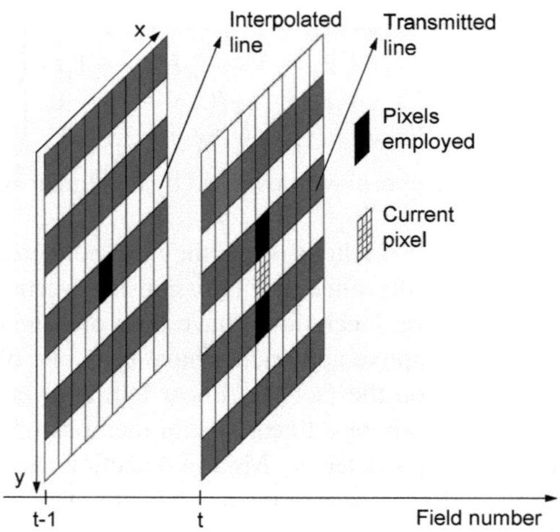

Fig. 1.21. Pixels employed by the median-based method in [13].

of vertical details and/or noise. An extensive study of this kind of algorithms is presented in Chapter 2.

1.3.1.2 Motion Compensated Methods

The first step to perform motion compensated algorithms is to calculate the motion vectors. In fact, the process to achieve it is independent of de-interlacing, and it is known as 'motion estimation' [20]. Initially, motion estimation algorithms repeat the calculations for every pixel in the image. However, since this process is very tedious, an efficient alternative is to estimate motion vectors on a block-by-block basis, where blocks are groups of N pixels. This kind of algorithms is called block-matching algorithms. Figure 1.22 illustrates the basic idea of block-matching algorithms. Motion trajectory is defined as the imaginary line that connects identical image parts in a sequence of images (the soccer ball in this example). The projection of this motion trajectory between two successive pictures on the image plane is called 'motion vector':

$$\mathbf{M} = (M_x \cdot \mathbf{u}_x, M_y \cdot \mathbf{u}_y) \tag{1.4}$$

where M_x and M_y are the coordinates of the motion vector in the horizontal and vertical directions, respectively.

As shown in Figure 1.22, the motion vector is assigned to the centre of a block in the current field by searching a similar block within a search area (blue square), also centered at the same spatial coordinates but in the previous field. To find the most

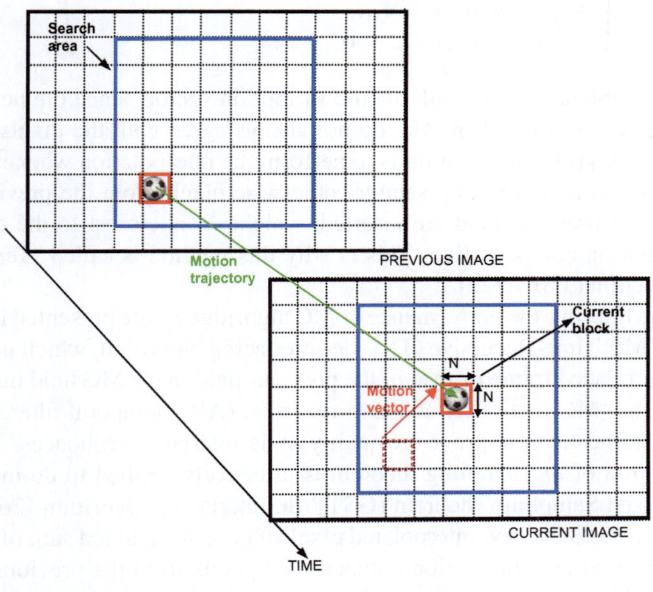

Fig. 1.22. Estimation of the motion vector by using block-matching algorithms.

suitable motion vector, a number of candidate vectors are evaluated by applying an error measure to quantify block similarity. The most popular error function is the 'Summed Absolute Difference (SAD)' criterion [20].

The process to search an efficient motion vector is not simple. Therefore, optimization strategies have been reported to find an optimal vector without an excessive number of calculations. The first motion vector estimator used in consumer products, called 3-D Recursive block matcher (3-D RS), reduces the search area in the direct spatio-temporal neighborhood [22]. The performance of 3-D RS algorithm has been compared with other algorithms that are available in consumer products, as well as algorithms from recent literature, in [20]. The results of these tests show that 3-D RS algorithm offers accurate motion vectors at a low complexity.

Once motion vectors are available, de-interlacing is performed along the motion trajectory. Motion compensation allows us to convert a moving sequence into a stationary one by replacing the pixels $I(x,y,t+m)$ with $I(x+m\cdot M_x, y+m\cdot M_y, t)$. For instance, MC field insertion method is performed as follows:

$$I_p(x,y,t) = I(x-M_x, y-M_y, t-1)) \tag{1.5}$$

In this way, the corresponding MC method of the VT-filtering proposal in equation (1.2) is calculated as follows:

$$I_p(x,y,t) = \left\{ \begin{array}{l} -\frac{1}{16}I(x-M_x, y-2-M_y, t-1) \\ +\frac{2}{16}I(x-M_x, y-M_y, t-1) \\ -\frac{1}{16}I(x-M_x, y+2-M_y, t-1) + \frac{8}{16}I(x, y-1, t) \\ +\frac{8}{16}I(x, y+1, t) - \frac{1}{16}I(x+M_x, y-2+M_y, t+1) \\ +\frac{2}{16}I(x+M_x, y+M_y, t+1) \\ -\frac{1}{16}I(x+M_x, y+2+M_y, t+1) \end{array} \right\} \tag{1.6}$$

Implicitly, a problem appears with the use of motion vectors since the new spatial coordinates $(x+m\cdot M_x, y+m\cdot M_y)$ do not always agree with the pixels in the interlaced sampling grid. One solution is to perform the interpolation whenever the motion vector points at an existing sample, otherwise pixels from the previous or the previous of the previous field are inserted, whichever is closest to the desired position after motion compensation. This is why this method is called 'Temporal Backward Projection (TBP)' [23].

Other ways to improve the performance of MC algorithms were presented in [24]. They were the 'MC Time-Recursive (TR)' de-interlacing algorithm, which uses the previously de-interlaced frame instead of the previous field in the MC field insertion algorithm, and the 'MC first-order adaptive-recursive (AR)' temporal filter, which uses adaptive parameters to improve the quality of de-interlaced sequences.

A generalization of the sampling theorem is effectively applied to de-interlace in the 'Generalized Sampling Theorem (GST)' de-interlacing algorithm [26]-[27]. This method calculates the new interpolated pixel value as a weighted sum of pixels from the current field and the motion compensated pixels from the previous field. The 'Robust GST' de-interlacing method in [28] increases the robustness of the GST algorithm by applying the median operator. Another improvement of the GST algorithm is presented in [29] as 'GST-2D'. In this case, the weighted sum is obtained after interpolating pixels of the previous, current, and next fields of the sequence. Its corresponding robust version has also been developed and named 'Robust GST-2D' de-interlacing algorithm.

1.4 Implementation of Video De-Interlacing

Since video material is captured until it is displayed, de-interlacing can be performed at various points. For instance, it can be done in the filming studios to achieve a better display resolution than the resolution of the captured image. Since time constraints are usually not critical in professional studios and quality is the objective, software implementations of powerful and costly de-interlacers are usually employed.

In the case of more modest studios or amateurs with standard computers, one option is the use of software programs for video post-production such as *AvySynth* for Microsoft Windows Operating System (OS) [30], *Avidemux* for either Microsoft Windows, Linux OS, and Mac OS [31], *VirtualDub* for Microsoft Windows [32], *DSCaler* for Microsoft Windows [33], and *Cinelerra* for Linux [34]. Amateur users can include plugings in either of these programs to perform several de-interlacing methods. A second option is the use of graphics cards. The development of consumer graphics cards during the last years has encouraged its use in the implementation of image processing algorithms. The improvement of their processing power has made that Graphics Processing Units (GPUs) achieve a remarkable performance when de-interlacing. For instance, the algorithm described in [35] has been implemented on common programmable graphics cards in [36]. Currently, Nvidia advertises PureVideo HD technology with all sorts of video processing that include the functionality of de-interlacing [37]. Nvidia has also employed this technology to achieve high quality video on mobile devices. On a PC desktop, the use of a video driver capable of de-interlacing is another alternative. One example is the video driver included in Windows Driver Kit for Windows Server 2008 and earlier versions of Microsoft Windows OS. Finally, the use of video players that have built-in a de-interlacing method is another solution on a PC desktop. Some examples are *InterVideo WinDVD Media Player* in Microsoft Windows OS [38], or *Quick Time Player* for Mac OS X that uses *JES Deinterlacer 3.3* [39], [40]. The algorithms included in these video players usually provide poor performance.

In consumer equipments, de-interlacing is usually performed on video processing chips. Figure 1.23 illustrates several consumer devices that require a progressive scanning: Liquid Crystal Display (LCD) TVs, plasma TVs, Audio-Video (AV) receivers, DVD players, High Definition (HD)-DVDs, Blu-ray players/recorders, and projectors. Independently of the video source (during the last years multiple video sources are proliferating), these consumer equipments contain video processing chips that normally perform intelligent and costly de-interlacing methods to obtain a high quality output progressive signal.

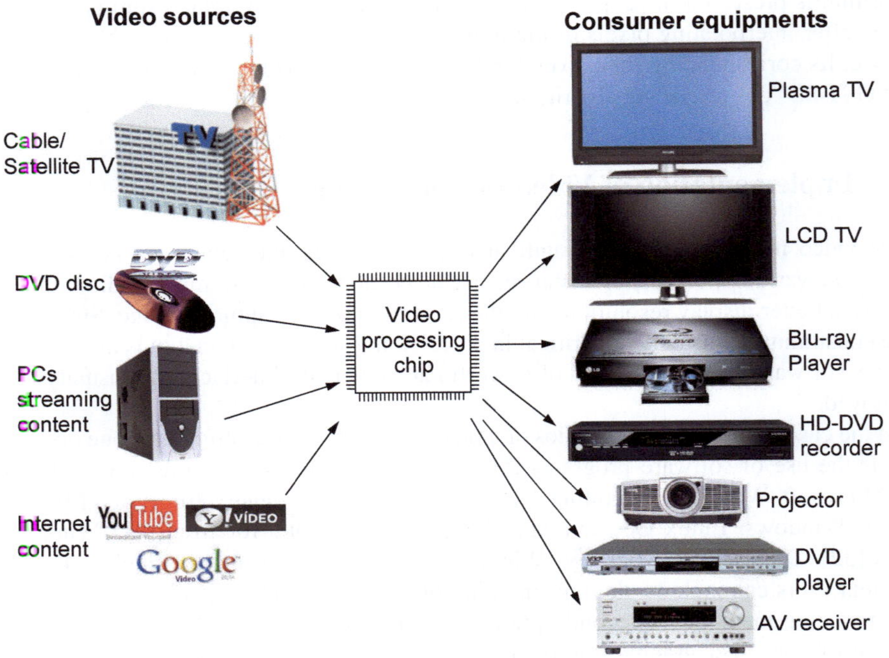

Fig. 1.23. Examples of equipments that contain video processing chips.

Several choices to implement these video processing chips such as Application-Specific Integrated Circuits (ASICs) [41]-[56], programmable solutions on Digital Signal Processing (DSPs) [57]-[61] and/or Field Programmable Gate Arrays (FPGAs) [64]-[65], and Intellectual Property (IP) cores [66]-[68] that can be used as building blocks within ASICs or FPGA designs are currently available in the market. Each of these alternatives offers advantages and disadvantages regarding high performance, flexibility and easy upgrade, and low development cost.

In commercial equipments, ASICs offer the most efficient solution justified by the huge demand of video products. According to Consumer Electronics Association (CEA) report, 26 billions dollars of a total of 155 billions of dollars in the industry of consumer electronics were destined to digital television [71]. Among this area, the development of semiconductor chips to implement video processing is the most important factor [72], [73]. In the other side, FPGAs are a good solution when there is a low volume of consumer products or in the case that the aim is to develop a prototype. Furthermore, FPGAs address with the requirement of flexibility and easiness to upgrade. Taking into account these considerations, FPGA implementation is the option chosen to implement the de-interlacing algorithms proposed in this dissertation.

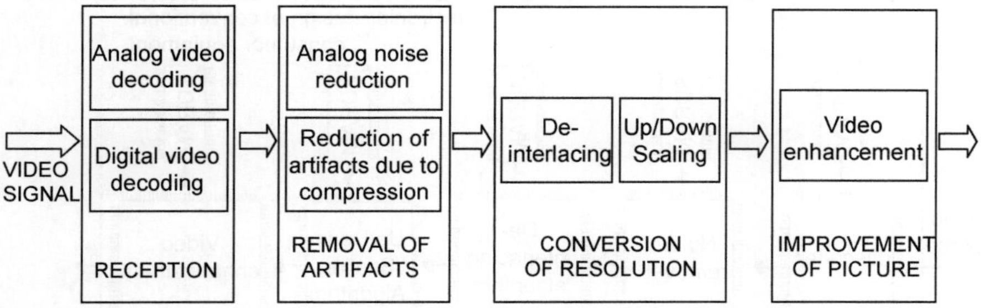

Fig. 1.24. The four stages of video format conversion.

1.4.1 Consumer Video Processing Chips

Video format conversion usually includes the four stages shown in the block diagram of Figure 1.24:

- Reception: video TV is received by a TV decoder, which is in charge of decoding the analog or digital signal. The TV decoder converts standard analog television signals in digital form according to ITU-R recommendations. In case that the signal is digital, the decoder deals with standards for video compression such as MPEG-2[2], H.264[3] or VC-1[4].
- Removal of artifacts: analog signal suffers white gaussian noise whereas digital signal is affected by video compression artifacts, which introduces two kinds of artifacts: ‘block’ and ‘mosquito’ noise.
- Conversion of resolution: signal is firstly converted from interlaced to progressive. After performing de-interlacing, an algorithm for down or up scaling is applied accordingly to the format required by the device.
- Improvement of picture: the quality of video signal is enhanced by applying several picture enhancement algorithms such as color improvement, sharpness or edge enhancement, and improvements of contrast, details and textures.

Although primitive consumer equipments had several ASICs to perform each of these four tasks, the current high-end products in the market contain a highly integrated chip to perform the last three tasks and even all the tasks as shown in Figure 1.25.

Since the volume of consumer products in the area of video processing is huge, numerous semiconductor companies offer series of ASICs for digital consumer applications. Although a generalization of the input/output signals of their products is

[2] MPEG-2 is widely used by digital TV transmission systems.

[3] H.264 (also known as MPEG-4 Part 10, or MPEG-4 AVC (for Advanced Video Coding)) is a standard for video compression. It is one of the latest block-oriented motion-estimation based codes developed by the ITU-T Video Coding Experts Group (VCEG) together with the ISO/IEC MPEG.

[4] This compression method is currently supported for HD-DVDs, Blu-ray Discs, and Windows Media Video 9.

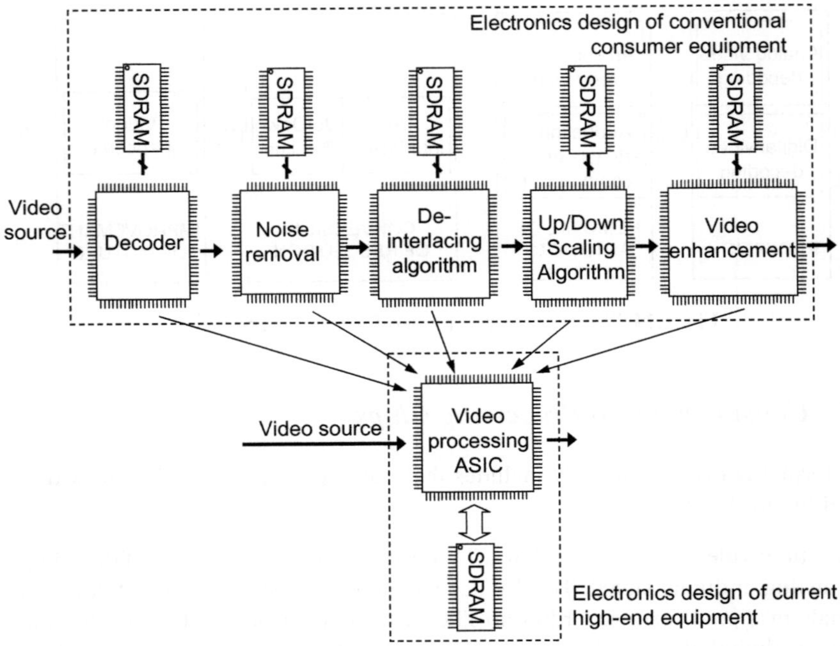

Fig. 1.25. The evolution of electronics design in consumer equipments.

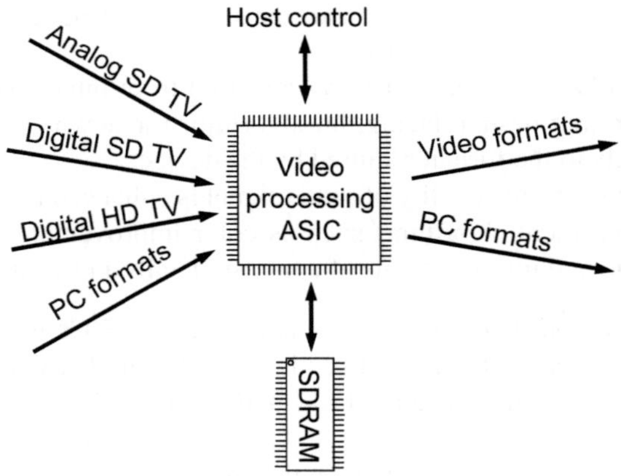

Fig. 1.26. Schematic illustration to show up input/output signals of a video processing ASIC.

difficult due to the high number of different devices, Figure 1.26 shows a schematic illustration of a video processing ASIC. The video formats at the output are usually 480i, 480p, 576i, 576p, 720p, 1080i, and 1080p, whereas the PC formats are VGA, SVGA, XGA, and SXGA.

The de-interlacing algorithms implemented on commercial video ASICs are not well detailed since they are protected by a high number of patents. However, the semiconductor companies advertise the main features of their algorithms. The following list summarizes the features offered by the main companies:

- Anchor Bay provides a video processing ASIC series called the ABTXXXX [41]. Among these chips, the ABT2010 provides a motion-adaptive and edge-adaptive de-interlacing. It uses five fields to detect motion and it is capable of dealing with hybrid material from film and video. One example of a commercial product that uses this chip is the Toshiba HD-DVD EP-30.
- The latest series of ASICs for digital consumer applications developed by NEC Electronics [42] and Silicon Optix [43] are based on Hollywood Quality Video (HQV) technology [44]. It performs a motion-adaptive de-interlacing by using four fields to detect motion. It also works with film and video material detection based on pixels. One example of a consumer product that includes one of these ASICs is the NC2500S digital cinema projector.
- Smartasic offers a series of ASICs called STVXXX [45]. Among them, STV102 performs a motion-adaptive de-interlacing, which is able to detect film mode. ViewSonic(R) Corporation utilizes an ASIC from Smartasic in its new ViewBox 50 HRTV.
- Trident Digital Media Inc. company offers a highly integrated System-On-a-Chip (SOC) device called SVP^{TM} [46] based on the so called DRCe technology. It performs an advanced motion-adaptive de-interlacing that supports film mode detection. Toshiba uses DCRe technology from Trident in the models of its Qosmio AV notebook, the Qosmio G15-AV501, and Qosmio F15-AV201.
- Among the ASICs offered by ST Microelectronics [47], SDTV 3550 performs de-interlacing by combining field merging, line interpolation, or motion-adaptive de-interlacing based on a median filter. In 2002, ST signed a five-year agreement with Thomson Multimedia to cooperate in the development of SOCs for DVD, digital TV, and digital set-top boxes.
- Genesis Microchip Inc. provides Directional Correlation De-interlacing (DCDi) technology [48] that dynamically detects edges in the image. One ASIC based on DCDi technology is the FLI2300, which is widely used by Samsung products, for instance in the HLN4365W HDTV.
- Silicon Image provides ASICs of SiI 90XX series for Plasma TVs, LCD TVs, and projectors [49]. SiI 8100 and SiI 504 are special ASICs for DVDs. All of them perform a motion-adaptive de-interlacing with detection of pull-down modes. SiI 1257 ASIC for Interlaced to Progressive Conversion (IPC) is used by some models of LG LCD TVs.
- Zoran Corporation offers ASICs for DTV products (supraHD family) and DVD products (Vaddis family) [50]. Both families are composed by highly integrated ASICs that implement motion-adaptive de-interlacing based on a pixel-by-pixel processing. One example of consumer products that uses Zoran technology is the HD DVD car stereo manufactured by Sony.
- Pixelworks offers the Digital Natural Expression (DNX) technology [51]. In the consumer market, Optoma projectors include ASICs based on DNX technology.

- Fujitsu provides Advanced Video Movement (AVM)-II technology, and its new version AVM-III since 2006 [52]. This technology is based on an image-adaptive processing with picture mode identification. It is included in full HD flat displays and projectors of Fujitsu AVIAMO series.
- Micronas offers a variety of ASICs in video signal processing [53]. One of them is the DPS 9455B, which is especially designed for LCD TVs and Plasma TVs. Micronas semiconductor serves several consumer brands worldwide.
- Philips products rely on NXP that offers Nexperia media processors [54]. They perform a high quality hardware image scaler and advanced de-interlacer, improved with media processing software to do motion compensation de-interlacing.
- Mstar Semiconductors offers ASICs for LCD displays such as MST9010 and MST9015 [55].
- Marvell provides Qdeo video processing technology [56]. It performs a film mode detection, which includes 3:2 and 2:2 pull-down processes but also other cadences such as 2-3-3-2, 6-4, etc. used in special cases. When no mode repetition is detected, motion and edge-adaptive de-interlacing is performed. LG BH200 HD DVD/Blu-ray player uses Marvell 88DE2710 digital video format converter with Qdeo video processing.

1.4.2 De-Interlacing Implementations Based on DSPs, FPGAs and IP Cores

Advances in programmable devices have enabled DSPs as a considerable solution to perform de-interlacing [57]-[61]. The purpose of the technical report in [57], developed by Texas Instruments company, is to show how the resizer hardware module in the video processing sub-system of the TMS320DM6446 DSP processor can be used for the purpose of de-interlacing. The implemented method is a simple spatial technique since it works with pixels in the rows of the neighborhood. In [58], a motion-adaptive de-interlacing technique which blends ELA and a temporal adaptive interpolation is tailored for Very Long Instruction Word (VLIW) DSP architecture. A motion compensation de-interlacing algorithm with film-detection is presented in [59]. The architecture to implement this algorithm is based on a software/hardware codesign. Film mode detection and motion estimation are implemented on a Trimedia CPU core from NXP to achieve a high de-interlacing quality level. Motion compensation de-interlacing is realized by specific hardware. In [60]-[61], complex motion compensation de-interlacing algorithms with film detection techniques are implemented in real-time on a Philips TriMedia TM1100 processor. The work presented in [60]-[61] analyzed up to seven different motion estimation algorithms to show up the good performance of the object based estimator approach developed on software.

Other authors have presented their de-interlacing implementations based on FPGAs. Some examples are described in [62]-[65]. In [62], a motion-adaptive de-interlacing method that uses fuzzy logic to detect motion is implemented in real-time on a FPGA. The algorithm proposed in [63] is synthesized through VHDL into FPGA components. It implements a motion-adaptive de-interlacing algorithm, which blends the 'Weave' method with the line average of pixels in the current and previous field and with an edge-adaptive spatial method. In [64], a non-linear filter algorithm prototyped on a Virtex FPGA is used to de-interlace broadcast television. The hardware architecture of a motion-adaptive de-interlacing algorithm with texture detection and its hardware implementation on a FPGA is presented in [65].

Several companies offer de-interlacing implementations based on IP cores [66]-[68]. Among them, Algolith Inc. [66] offers an IP core called MADI that performs a motion-adaptive de-interlacing and two more IP cores that increase the quality of the de-interlaced signal: CINETEL and VOFD (Video Over Film Detection) IP cores to detect film and hybrid material, respectively. VOFD has to be used in conjunction with CINETEL since it looks for video material when the picture is classified as film material. Recently, Algolith Inc. and Altera have announced the integration of IP cores designed by Algolith Inc. with Altera's Cyclone FPGAs [67]. Moreover, Altera delivers intellectual property (IP) cores in its MegaCore IP Library Evaluation. Among them, Deinterlacer IP core supports four de-interlacing methods: conventional 'Bob', 'Bob' with line average of pixels in the upper and lower lines, 'Weave', and a simple motion-adaptive algorithm.

Image Processing Techniques Ltd. together with Xilinx offers Advanced Video Development Platform (AVDP) [68]. This company provides an IP core to perform motion-adaptive de-interlacing on the AVDP platform. The level of motion is evaluated by measuring the luminance and chroma differences between pixels in successive frames on a pixel-by-pixel processing.

Zoran company provides an IP core for video de-interlacing called Frame-it-1 [69]. It performs a motion-adaptive de-interlacing capable of detecting material originated from film. Zoran's IP core for video de-interlacing is especially designed for video ASIC design. Other semiconductor company that offers IP cores for ASIC design is Mstar Semiconductors [70][5].

1.5 The Role of Fuzzy Logic in Video Processing

The successful achieved by fuzzy systems in many application fields is due to the ability of fuzzy logic to model uncertainty and subjective concepts in a better form than conventional methods. In the area of video processing it is common to deal with concepts like brightness, edges, motion, uniformity, etc., which contain certain amount of uncertainty. For instance, it is difficult to define if a pixel belongs to an edge (i.e., object frontier) or not. Especially this decision is not trivial when the images are noisy and/or contain a high number of details. Fuzzy logic theory provides a mathematical framework to deal with this kind of uncertain information [74]. In this sense, several proposals have been presented in the literature to extract edges without being deceived by the noise which is presented in the image [75]-[77]. Detection of edges in an image is a very important step in order to realize a complete image understanding system. In fact, edges often correspond to object

[5] This manufacturer does not provide information about which kind of de-interlacing algorithm is performed.

boundaries, shadow transitions, and other image details or features, and this is why a correct edge detection is crucial.

Furthermore, fuzzy logic-based systems are well suited to emulate the approximate reasoning mechanisms used by the human brain. A fuzzy system acquires and represents its knowledge base symbolically, while knowledge is processed numerically. The first feature allows fuzzy logic-based systems to model knowledge from natural language, and organize such knowledge into fuzzy logic inference rules [78]. The second feature allows the use of fast numerical algorithms and the direct implementation of the systems. In certain tasks of video processing, the ability of fuzzy logic-based systems to emulate human reasoning is very useful. For instance, fuzzy logic-based systems have widely been employed to perform image enhancement, that is, to improve the subjective picture quality [79]. This includes the reduction of noise, the de-blurring of edges, the highlighting of specified features, and the re-mapping of luminance values in order to obtain an improved contrast or color [80]-[102].

Many authors have demonstrated that fuzzy systems are universal approximators [103]-[105]. In general, they provide a non-linear mapping between the input and the output space that is adjusted by properly choosing the fuzzy system parameters. This universal approximation capability has been exploited to interpolate images [110]-[112] and sequences of images [113]-[115]. In the case of video sequences, several approaches have been proposed for de-interlacing purpose. In particular, the proposal in [113] uses soft-decision fuzzy logic techniques to remove noise and de-interlace TV video signals. In [114], a fuzzy edge-direction detector is introduced to orient a conventional vertico-temporal filter for de-interlacing tasks. The proposal in [115] uses fuzzy inference rule bases to detect motion and to choose the de-interlacing strategy according to the presence of motion detected. This approach is described in detail in Chapter 2.

To understand better the advantages of fuzzy modeling over crisp-based algorithms, a brief description of fuzzy basic concepts is introduced in subsection 1.6.1. In subsection 1.6.2, CAD environments for designing fuzzy systems are summarized.

1.5.1 Basic Concepts of Fuzzy Logic Theory

A fuzzy inference system is composed by a set of rules and an inference mechanism. The rules use linguistic terms such as *small* and *large* that are usually employed in natural language by humans. The inference mechanism provides correct conclusions from incomplete or uncertain information.

1.5.1.1 Fuzzy Sets and Membership Functions

Fuzzy set is the fundamental concept in which the fuzzy logic theory is based. From a formal point of view, the concept of fuzzy set is a generalization of the classical concept of set. The fundamental difference lies in that, while in the classical set theory a certain element can be a member of a set or not, in the fuzzy set theory proposed in [116], an element can be a member of several sets with different degrees of

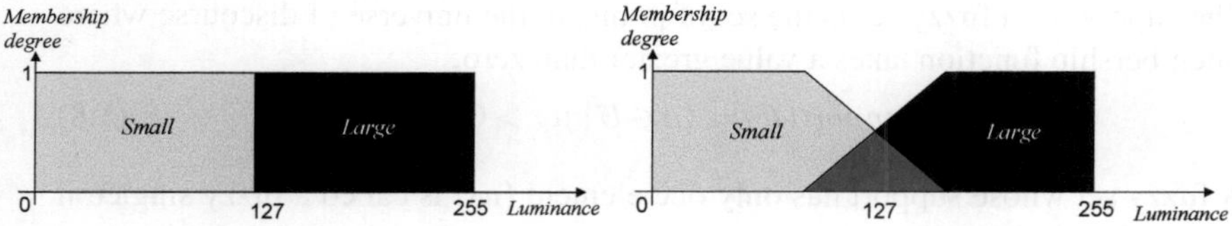

(a) Classical sets (b) Fuzzy sets

Fig. 1.27. Graphical representation of classical and fuzzy sets.

membership. Figure 1.27 graphically compares the concepts of classical sets 1.27(a) and fuzzy sets 1.27(b).

Mathematically, a fuzzy set can be represented by a set of ordered pairs that assign a membership degree for each element in the universe of discourse U:

$$F = \{(u, \mu_F(u) \mid u \,\varepsilon\, U\} \tag{1.7}$$

The membership degree $\mu_F(u)$ depends on the membership function. It takes the values in the interval [0,1]. The selection of the shape for a membership function is subjective and it depends on the context. Nevertheless, the most employed functions are shown in Figure 1.28.

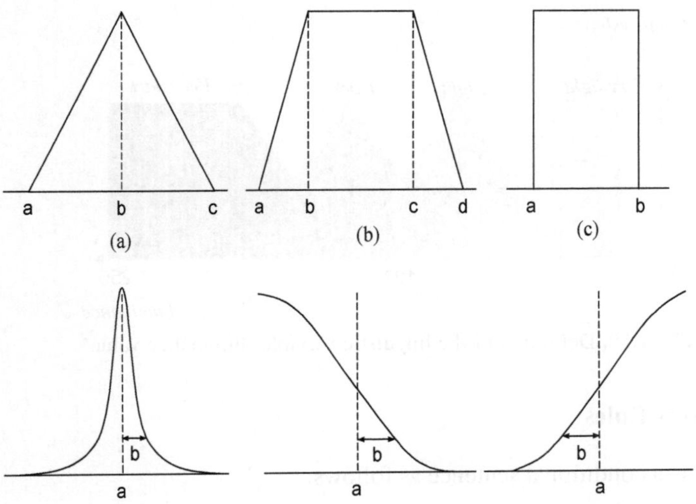

Fig. 1.28. The most common membership function shapes: (a) triangular, (b) trapezoidal, (c) rectangular, (d) gaussian, (e) Z-function, (f) S-function.

The support of a fuzzy set is the set of points in the universe of discourse where the membership function takes a value greater than zero:

$$support(F) = \{u \in U | \ \mu_F > 0\} \tag{1.8}$$

A fuzzy set whose support has only one element (u_0) is called a fuzzy singleton (or fuzzy singularity):

$$\mu_F(u) = 0 \ if \ u \neq u_0, \ \mu_F(u_0) = 1 \tag{1.9}$$

1.5.1.2 Linguistic Variables

A linguistic variable is a variable whose values are expressed by means of natural language terms. The different terms or linguistic values are represented with fuzzy sets characterized by membership functions defined on the universe of discourse. A linguistic variable is characterized by a quadruple of elements (X, T, U, M), where X is the name assigned to the linguistic variable, T is the set of linguistic labels or values the variable can take, U is the application domain or universe of discourse where the variable is defined, and M is a semantic rule associating each linguistic label with a fuzzy set defined on the universe of discourse U.

Figure 1.29 shows an example where the linguistic variable is the *'luminance value'*, the set of linguistic values are *'very light, light, dark, very dark'*, the universe of discourse is the range 0-255, and the meaning of the linguistic labels is defined by the shown membership functions.

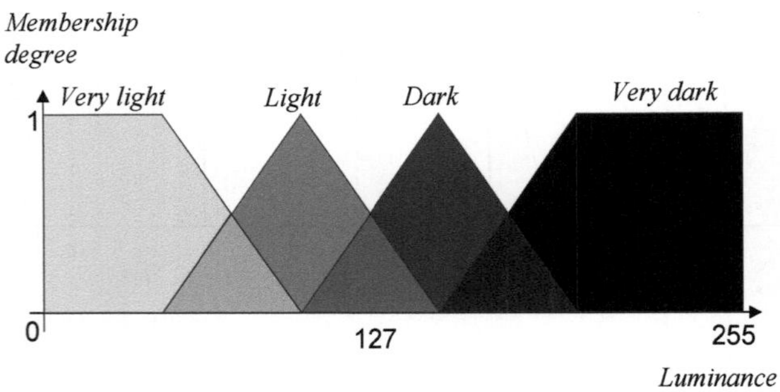

Fig. 1.29. Definition of the linguistic variable 'luminance value'.

1.5.1.3 Fuzzy Rules

A fuzzy rule is a conditional sentence as follows:

$$if \ \{fuzzy \ proposition\} \ then \ \{fuzzy \ proposition\} \tag{1.10}$$

where the proposition on the left side is called the rule antecedent, and the one on the right side is called the rule consequent.

An atomic fuzzy proposition is a simple sentence associating a linguistic label to a linguistic variable. A compound fuzzy proposition is any combination of atomic fuzzy propositions, employing linguistic connectives such as conjunctive (*'and'*), disjunctive (*'or'*), or negations (*'not'*). For instance, a compound fuzzy proposition to smooth the impulse noise in an image could be: *if the luminance pixel value in the upper line is dark and the luminance pixel value in the lower line is dark then the luminance value of the current pixel is dark.*

In order to calculate the membership function of a compound proposition, it is necessary to define the interpretation of the linguistic connectives *'and'*, *'or'*, and the operator *'not'*.

- Union: the union of two fuzzy sets A and B is a fuzzy set whose membership function for a certain element (u) of the universe of discourse U is given by the following expression:

$$\mu_{A\cup B}(u) = Snorm\left[\mu_A(u), \mu_B(u)\right], \ \forall u \in U \tag{1.11}$$

  where *Snorm* is an operator such as maximum, bounded sum, etc., that represents the linguistic connective 'or'.
- Intersection: the intersection of two fuzzy sets A and B is a fuzzy set whose membership function for a certain element (u) of the universe of discourse U is given by the following expression:

$$\mu_{A\cap B}(u) = Tnorm\left[\mu_A(u), \mu_B(u)\right], \ \forall u \in U \tag{1.12}$$

  where *Tnorm* is an operator such as minimum, product, etc., that represents the linguistic connective 'and'.
- Complement: the complement of a fuzzy set A is a fuzzy set $\bar{A}$ with usually the following membership function:

$$\mu_{\bar{A}}(u) = 1 - \mu_A(u), \ \forall u \in U \tag{1.13}$$

The fuzzy implication operator or implication function, ϕ, expresses the fuzzy relation between the antecedent and the consequent of the rule *'if u is A then v is B'*, as follows:

$$\mu_{A\to B}(u,v) = \phi \ \left[\mu_A(u), \mu_B(v)\right] \tag{1.14}$$

Several proposals of implication functions, which use different options for performing the operations of the set intersection, union, and complement, have been presented in the literature. Most of them are extensions of the implications used in propositional or multivalued logic [117], [118].

1.5.1.4 Approximate Reasoning Techniques

Reasoning techniques permit obtaining logical deductions, that is, inferring conclusions from premises. The propositions of classical logic admit only two truth values {'0' and '1'}. However, in fuzzy logic the truth value of a proposition can be any number in the $[0,1]$ interval. This generalization is the basis for the approximate reasoning techniques that permit obtaining right conclusions from imprecise and vague premises.

Table 1.2. Inference mechanisms of the classical logic

	Modus ponens	Modus tollens
Rule	*if u is A* $\rightarrow$ *v is B*	*if u is A* $\rightarrow$ *v is B*
Premise	*u is A*	*v is not B*
Conclusion	*v is B*	*u is not A*

The two most important inference rules in propositional logic are called *modus ponens* and *modus tollens*. Modus ponens states that, given the rule *'if u is A then v is B'* and the observation *'u is A'*, the conclusion *'v is B'* is obtained. Modus ponens is used to obtain conclusions by means of forward inference, and it is the most widely used in control applications where there is a relationship between cause and effect.

The second inference rule, modus tollens, states that, given the rule *'if u is A then v is B'* and the observation *'v is not B'*, the conclusion *'u is not A'* is obtained. Modus tollens is employed to deduce causes by means of backward inference. Table 1.2 summarizes these inference rules.

These two inference rules are extended in fuzzy logic as it is shown in Table 1.3, where {*A, B, A', B'*} are linguistic terms represented by fuzzy sets, with the feature that *B'* will approximate more *B* as *A'* approximates more *A*. In the case of a fuzzy modus tollens *A'* will be more different from *A* as *B'* is more different from *B*.

Focusing on the case of generalized modus ponens, the problem is the numerical evaluation of *B'* from the similarity degree between *A* and *A'*. Zadeh gave a solution to this problem when he proposed the *compositional rule of inference* stating that, given a rule and an observation (*A'*), the conclusion *B'* can be obtained as the sup-star composition ($\circ$) between the observation and the fuzzy relation, R, induced by the rule:

$$B' = A' \circ R \tag{1.15}$$

Applying the definition of sup-star composition, the membership function for the fuzzy set obtained as conclusion is:

$$\mu_{B'}(v) = \forall_U \{\mu_{A'}(u) * \mu_{A\rightarrow B}(u,v)\} \tag{1.16}$$

where $\forall$ symbolizes the supreme operator extended to all the elements of the universe of discourse U, and $*$ is a $Tnorm$. Substituting in (1.16) the membership function of the fuzzy relation corresponding to the implication function defined by (1.14), the membership function of the fuzzy set B' can be expressed as:

$$\mu_{B'}(v) = \forall_U \{\mu_{A'}(u) * [\phi\ [\mu_A(u), \mu_B(v)]]\} \tag{1.17}$$

Table 1.3. Inference mechanisms of the fuzzy logic

	Modus ponens	Modus tollens
Rule	*if u is A* $\rightarrow$ *v is B*	*if u is A* $\rightarrow$ *v is B*
Premise	*u is A'*	*v is not B'*
Conclusion	*v is B'*	*u is not A'*

Therefore, the result of the inference depends on the choices for the operator * and the implication function ϕ. With respect to the $Tnorm$ used in the compositional rule of inference, Zadeh initially proposed the use of a minimum (composition sup-min) [119]. The sup-min and sup-product composition methods are, undoubtedly, the most widely employed. In many applications, for instance in control and many image and video processing applications [117], the premises are of the singleton type '*u is* u_o', where u_o is a fuzzy singleton. In these cases, the expression in (1.17) is given by the expression:

$$\mu_{B'}(v) = \phi\left[\mu_A(u_o), \mu_B(v)\right] \tag{1.18}$$

1.5.1.5 Rule-Based Inference Mechanisms

The compositional rule of inference provides a mechanism for calculating the result of a fuzzy rule. However, the rule base of a fuzzy inference system normally contains a set of linguistic description rules. Let us consider a fuzzy inference system with multiple inputs and a single output (MISO system). The antecedents of the rules can contain any combination of the inputs by means of the conjunction operators *'and' 'or'*. In contrast to a conventional expert system, a fuzzy system can have several active rules simultaneously. The global conclusion must be calculated from the aggregation of the partial solutions supplied by each rule. The choice of different options to implement connectives, implication functions, and aggregation operators gives rise to a set of well-known inference mechanisms.

The Mamdani's inference mechanism uses the minimum as implication operator and the maximum as aggregation operator for each partial rule conclusion. The membership function of the conclusion inferred from the application of the rule base is given by the following expression:

$$\mu_{B'}(v) = max_r(min(h^r, \mu_B^r(v))) \tag{1.19}$$

where h^r is the activation degree of the r-th rule.

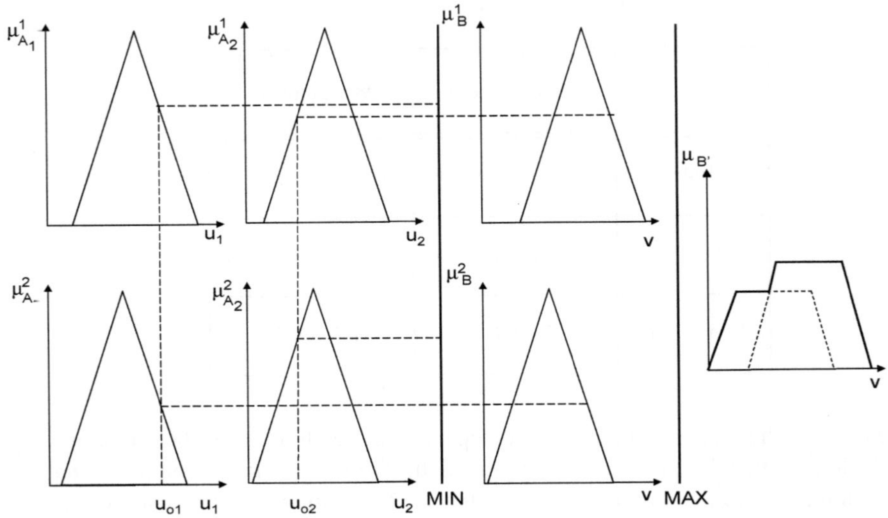

Fig. 1.30. Min-max or Mamdani's inference mechanism.

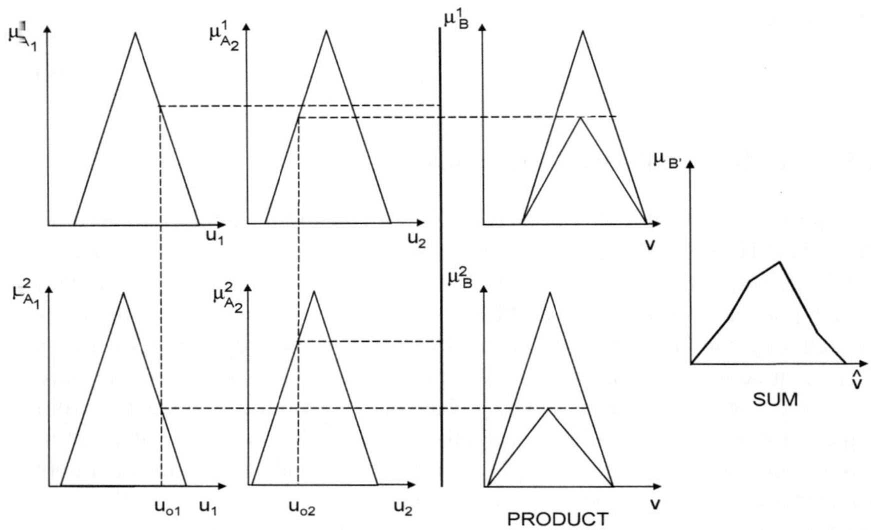

Fig. 1.31. Product-sum inference mechanism.

If inputs of the system are singletons ($u = (u_{o1}, u_{o2}, \ldots, u_{on})$) and the connectives of the antecedents are conjunctive, the activation degree of the r-th rule is given by the following expression:

$$h^r = min(\mu^r_{A1}(u_{o1}), \mu^r_{A2}(u_{o2}), \ldots, \mu^r_{An}(u_{on}) \tag{1.20}$$

Figure 1.30 shows the graphical interpretation of this inference mechanism when the rule base contains two rules with two antecedents and one consequent.

The product-sum mechanism considers the product as implication operator and the sum as aggregation operator of each partial rule conclusion. The membership function of the resulting fuzzy set is given by:

$$\mu_{B'}(v) = \sum h^r \mu^r_B(v) \tag{1.21}$$

Figure 1.31 shows the graphical interpretation of the 'product-sum' inference mechanism when the rule base contains two rules with two antecedents and one consequent.

1.5.1.6 Defuzzification Methods

The result of any of the above commented inference mechanisms is a fuzzy set. For this information to be used in certain applications, like control and many image and video processing, it is necessary to procure a precise (crisp) value representing this set. The goal of the defuzzification methods is to provide this value. A great number of defuzzification methods have been proposed in the literature [120]-[122].

Among the traditional methods, one of the most commonly used in Mamdani type systems is the called Center of Area (CoA) or Center of Gravity (CoG). This method determines the representative value for a fuzzy inference system conclusion as the center of the area delimited by the fuzzy set *B'* resulting from applying the different rules. Practical applications employ the discrete version of this method:

$$CoA \rightarrow \hat{v} = \frac{\sum_{i=1}^{n} v_i \cdot \mu_{B'}(v_i)}{\sum_{i=1}^{n} \mu_{B'}(v_i)} \tag{1.22}$$

where n is the number of elements in the output universe of discourse.

The main drawback of this method, from the implementation point of view, is the need of sweeping the whole universe of discourse. More efficient implementations discard the handling of fuzzy information in rule consequents and substitute this information by a set of parameters representing it [120]-[122]. The resulting methods are known as simplified defuzzification methods. Each one of these methods differs from the others in the number and meaning of the parameters it uses.

Among simplified defuzzification methods, one of the most widely employed is the Fuzzy Mean (FM) or Height Method (HM), which consists of choosing as consequent of each rule the element with maximal membership. The system output is the weighted average of these crisp consequents, c^r, where the weights, h^r, are the activation degrees:

$$FM/HM \rightarrow \hat{v} = \frac{\sum_r h^r \cdot c^r}{\sum_r h^r} \tag{1.23}$$

Other simplified methods introduce a second parameter (ω) to consider the area or the support of the output fuzzy sets. Among them, the Weighted Fuzzy Mean (WFM), Jager's method (YM) or the Center of Sum Minimum (CoSM) are well known[122]. In these cases, the output value can be calculated by the expression:

$$WFM/YM/CoSm \rightarrow \hat{v} = \frac{\sum_r h^r \cdot c^r \cdot \omega^r}{\sum_r h^r \cdot \omega^r} \tag{1.24}$$

1.5.1.7 Takagi-Sugeno-Type Fuzzy Systems

The inference systems previously commented use consequents described by fuzzy sets. These types of fuzzy systems are usually referred to as *Mamdani-type systems*. Another type of fuzzy systems that have been widely employed are named *Takagi-Sugeno-type systems* [120]. They are characterized by the use of functions of their input variables as consequents [123]:

$$F_B^r(u_1, u_2, \dots u_n) = \sum_k a_k^r \cdot u_k + a_o^r \tag{1.25}$$

where $a_k^r (k = 1, 2, \dots, n)$ and a_o^r are constants.

The partial conclusions provided by the different rules coincide with the values that those functions take for a certain value of the inputs, u_o. To obtain the global output, each rule is weighted by its activation degree h^r:

$$v_o = \frac{\sum_r h^r \cdot F_B^r(u)}{\sum_r h^r} \tag{1.26}$$

Zero-order Takagi-Sugeno-type systems, in which $F_B^r(u) = a_o^r$, have widely been employed because they allow very simple software as well as hardware implementations. Since the consequents are represented by singleton values, they are also known as *singleton fuzzy systems*. A Mamdani-type system which employs the FM or HM as defuzzification method is equivalent to a zero-order Takagi-Sugeno type or a singleton fuzzy system that employs the centroids of the consequent fuzzy sets as singletons values for the consequents.

1.5.2 CAD Tools for Designing Fuzzy Systems

Although several fuzzy logic design environments are currently available [124]-[125], most of them are tailored for control applications and are not suitable for image or video processing. Only a modest tool has been reported in the literature that

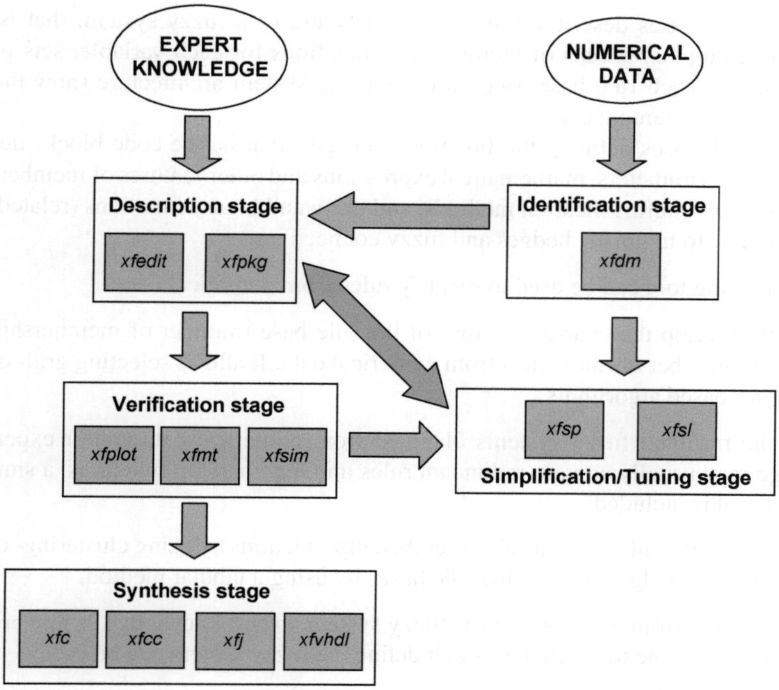

Fig. 1.32. CAD tools included in Xfuzzy 3.1.

provides a graphical interface to design fuzzy knowledge bases to process images [126].

In the research work described herein the capabilities offered by the environment Xfuzzy (version Xfuzzy 3.1) have been exploited to design fuzzy systems for video processing. Xfuzzy environment has been developed by our research group to ease the design of fuzzy systems by starting from linguistic and/or numerical knowledge up to final implementing them as hardware and/or software modules. The different versions of Xfuzzy have been distributed freely under GNU General Public License [127]. The latest version of Xfuzzy, called Xfuzzy 3.1, is entirely programmed in Java. It includes a wide set of new featured tools that allow automating the whole design process of a fuzzy logic-based system: from its description by using a formal specification language (the XFL3 language) to its synthesis in C, C++ or Java (to be included in software projects) or in VHDL (for hardware projects). Xfuzzy 3.1 contains a set of CAD tools which offer Graphical User Interfaces to ease the design flow at the stages of description, verification, extraction or identification of fuzzy rules from numerical data, tuning, simplification of fuzzy rule bases, and synthesis (see Figure 1.32) [128]-[129]. The tools that can be used at the description stage are the following:

- *xfedit*, which eases describing the logical structure of a fuzzy system, that is, its inputs, outputs, groups of memberships functions for each variable, sets of operators for each rule base, rule bases, and the system architecture (how the rule bases are interconnected).
- *xfpkg*, which eases defining the function packages, that is, the code blocks describing the parameters, mathematical expressions and other features of membership functions, defuzzification methods, and unity and binary functions (related, respectively, to linguistic hedges and fuzzy connectives.)

The following tool can be used to identify rules from numerical data:

- *xfdm*, to develop the coarse structure of the rule base (number of membership functions, number of rules, etc.) from numerical data. It allows selecting grid- or clustering-based algorithms.

Since the resultant fuzzy systems obtained from numerical data and/or expert knowledge may usually contain redundant rules and membership functions, a simplification tool is included:

- *xfsp*, to simplify either the variable membership functions by using clustering- or similarity-based algorithms or the rule bases by using a tabular method.

Independently from the origin of the fuzzy system, expert knowledge or numerical data, a tool to tune the parameters that define the fuzzy system can be used:

- *xfsl*, which allows applying a wide set of supervised learning algorithms (gradient-descent, second-order,Gauss-Newton, and statistical algorithms).

To verify the correct behavior of a fuzzy system, the following tools are included:

- *xfplot*, to visualize graphically one of the outputs of the system against 1 or 2 of its inputs.
- *xfmt*, to monitor how the output values are inferred from the input ones.
- *xfsim*, to simulate how the fuzzy system behaves within the application domain.

To synthesize the final fuzzy system, the following tools can be employed:

- *xfc, xfcc, xfj, and xfvhdl*, which translate the description of the system in XFL3 to C, C++, Java, and VHDL code, respectively.

The tools that have been particularly used for the purposes of our research have been *xfedit*, *xfpkg*, and *xfsl*. Although specific Java classes have been developed to simulate our fuzzy systems in image processing applications with the tool *xfsim*, a great part of the simulation work has been done by using the Image Processing Toolbox of Matlab.

1.6 Conclusions

De-interlacing plays a key role in video processing because analog and some digital transmission systems use an interlaced scanning whereas modern display devices require a progressive format. In addition, de-interlacing is the first step to be implemented in other video processing tasks such as picture conversion between formats. This is why research on de-interlacing algorithms is and has been so active.

Three features in the images have been identified as key points to achieve a good de-interlacing: motion, edges, and possible repetition of areas of fields. Evaluation of how much these features are present in an image and how they should influence interpolation decisions could better be done in a fuzzy rather than a crisp way. Validation of this hypothesis is the general objective of this book. This is why the following chapters focus on how to design efficient de-interlacers taking into account fuzzy considerations about motion, edges, and picture repetition.

Hardware implementation of de-interlacing algorithms is demanded especially in the field of consumer electronics. As a matter of fact, de-interlacing is one of the basic tasks implemented by any video processing chip available in the market. In addition, several approaches have been reported to implement de-interlacing algorithms on specific processors, such as DSPs and graphics processors, as well as on FPGAs.

References

1. Whitaker, J.: Television transmissions systems. In: Standard handbook of video and television engineering. McGraw-Hill Editorial, Blacklick OH (2002)
2. Russ, J.C.: The image processing handbook, 4th edn. CRC Press Editorial, Florida (2002)
3. Web page address, `http://www.itu.int`
4. Hart, J.A.: Technology, television and competition. Cambridge University Press, Cambridge (2004)
5. Whitaker, J.: Video signal measurement and analysis. In: Standard handbook of video and television engineering. McGraw-Hill Editorial, Blacklick OH (2002)
6. Symes, P.: Video compression demystified. McGraw-Hill Editorial, Blacklick OH (2001)
7. Whitaker, J.: DTV Handbook. The revolution in digital video. McGraw-Hill Editorial, Blacklick OH (2001)
8. Robin, M., Poulin, M.: Digital television fundamentals. McGraw-Hill Editorial, New York (1998)
9. de Haan, G.: De-interlacing. In: Digital Video. Post Processing, September 2006, pp. 185–201. University Press Eindhoven (2006)
10. Doyle, T., Looymans, M.: Progressive scan conversion using edge information. In: Chiariglione, L. (ed.) Signal Processing of HDTV, II, pp. 711–721. Elsevier Science Publishers, Amsterdam (1990)
11. Preliminary data sheet of Genesis gmVLDS, 8-bit Digital Video Line Doubler. version 1.0 (June 1996)
12. Weston, M.: Interpolating lines of video signals. Assignee: British Broadcasting Corporation, London, GB2. United States Patent Office US 4,789,893, December 6 (1988)
13. Haavisto, P., Juhola, J., Neuvo, Y.: Fractional frame rete up-conversion using weighted median filters. IEEE Trans. on Consumer Electronics 35(3), 272–278 (1989)
14. Haavisto, P., Neuvo, Y.: Motion adaptive scan rate up-conversion. Multidimensional Systems Signal Processing (3), 113–130 (1992)

15. Bock, A.M.: Motion-adaptive standards conversion between formats of similar field rates. Signal Processing: Image Communication 6(3), 275–280 (1994)
16. Jiang, H., Huu, D., Tinyork, E.: Motion adaptive deinterlacing. United States Patent (US 6,459,455) (October 2002)
17. Lin, W.-K., Lu, C.-Y.: Method for motion pixel detection. United States Patent (US 7,034,888) (April 2006)
18. Lin, W.-K., Lu, C.-Y.: Method for motion pixel detection with adaptive thresholds. United States Patent (US 7,242,435) (July 2007)
19. Philips Semiconductors, Datasheet SAA4990, PROZONIC (1995)
20. de Haan, G.: Motion estimation. In: Video Processing, pp. 221–269. University Press Eindhoven (2004)
21. de Haan, G., Biezen, P.: Sub-pixel motion estimation with 3-D recursive search block-matching. Signal Processing: Image Communications 6, 229–239 (1994)
22. de Haan, G., Biezen, P., Huijgen, H., Ojo, O.A.: True motion estimation with 3-D recursive search block-matching. IEEE Trans. on Circuits and Systems for Video Technology 3(5), 368–388 (1993)
23. Woods, J.W., Han, S.-C.: Hierarchical motion compensated de-interlacing. In: Proc. of the SPIE, vol. 1605, pp. 805–810 (1991)
24. Wang, F., Anastassiou, D., Netravali, A.: Time-recursive de-interlacing for IDTV and pyramid coding. Signal Processing: Image Communications 2, 361–366 (1993)
25. Yen, J.L.: On nonuniform sampling of bandwidth-limited signals. IRE Trans. on Circuit Theory, CT-3 3, 251–257 (1956)
26. Delogne, P., Cuvelier, L., Maison, B., Van Caillie, B., Vandendorpe, L.: Improved interpolation, motion estimation and compensation for interlaced pictures. IEEE Trans. on Image Processing 3(5), 482–491 (1994)
27. Vandendorpe, L., Cuvelier, L., Maison, B., Quelez, P., Delogne, P.: Motion-compensated conversion from interlaced to progressive formats. Signal Processing: Image Communication 6, 193–211 (1994)
28. Bellers, E.B., de Haan, G.: Advanced motion estimation and motion compensated de-interlacing. SMPTE Journal 106(11), 777–786 (1997)
29. Ciuhu, C., de Haan, G.: A two-dimensional generalized sampling theory and application to de-interlacing. In: Proc. of VCIP, January 2004, pp. 700–711 (2004)
30. `http://avisynth.org`
31. `http://avidemux.org`
32. `http://www.virtualdub.org`
33. `http://www.dscaler.org`
34. `http://cinelerra.org`
35. Gutiérrez-Ríos, J., Fernández-Hernández, F., Crespo, J.C., Trevino, G.: Motion adaptive fuzzy video de-interlacing method based on convolution techniques. In: Proc. of Information Processing and Management of Uncertainty in Knowledge-Based Systems (IPMU), Perugia (Italy) (July 2004)
36. Montemayor, A.S., Fernández, F., Gutiérrez, J., Cabido, R.: Fuzzy motion detector based on 2D-convolutions for deinterlacing using graphics cards. In: Proc. 11th Information Processing and Management of Uncertainty in Knowledge-Based Systems (IPMU) (July 2006)
37. `http://www.nvidia.es/page/purevideo`
38. `http://apps.corel.com/lp/ivi`
39. `http://www.apple.com/quicktime`
40. `http://www.macupdate.com/info.php/id/13069/jes-deinterlacer`
41. `http://www.anchorbaytech.com`

42. http://www.necel.com
43. http://www.siliconoptix.com
44. http://www.hqv.com
45. http://www.smartasic.com
46. http://www.tridentmicro.com
47. http://www.st.com
48. http://www.genesismicrochip.com/products
49. http://www.siimage.com
50. http://www.zoran.com
51. http://pixelworks.com/dnx
52. http://www.fujitsu-general.com
53. http://www.micronas.com
54. http://www.nxp.com
55. http://www.mstarsemi.com
56. http://www.marvell.com
57. He, Z., Natarajan, S.: De-interlacing and YUV 4:2:2 to 4:2:0 conversion on TMS320DM6446 using the resizer. Application report from Texas Instruments, pp. 1–17 (February 2008)
58. Peng, C., He, Z., Cager, Y.: An efficient motion-adaption de-interlacing technique on VLIW DSP architecture. In: IEEE Conference on Advanced Video and Signal Based Surveillance (AVSS), September 2005, pp. 644–649 (2005)
59. Bellers, E.B., Janssen, J.G.W., Schutten, R.J.: A software approach to high-quality video format conversion. In: Int. Conference on Consumer Electronics (ICCE), January 2005, pp. 441–442 (2005)
60. Wittebrood, R.B., de Haan, G.: Second generation video format conversion software for a digital signal processor. IEEE Trans. on Consumer Electronics 46(3), 857–865 (2000)
61. Wittebrood, R.B., de Haan, G.: Second generation DSP software for picture rate conversion. In: Int. Conference on Consumer Electronics (ICCE), June 2000, pp. 230–231 (2000)
62. Van de Ville, D., Van de Wall, R., Philips, W., Lemahieu, I.: Motion adaptive de-interlacing using fuzzy logic. In: Proc. International Conference on Information Processing and Management of Uncertainty in Knowledge-Based Systems (IPMU), July 2002, pp. 1989–1996 (2002)
63. Simonetti, R., Filisan, A.P., Carrato, S., Ramponi, G., Sicuranza, G.: A deinterlacer for IQTV receivers and multimedia applications. IEEE Trans. on Consumer Electronics 39(3), 234–240 (1993)
64. McWillians, P.J., McLaughlin, S., Laurenson, D.I., Collis, W.B., Weston, M., White, P.R.: Non-linear filtering for broadcast television: a real-time FPGA implementation. In: Proc. IEEE Int. Conf. Image Processing (ICIP), Thessaloniki (Greece), October 2001, vol. 3, pp. 354–357 (2001)
65. Yuhong, Y., Yingqi, C., Wenjun, Z.: Motion adaptive deinterlacing combining with texture detection and its hardware FPGA implementation. In: Proc. IEEE Int. Workshop VLSI Design and Video Technology, Suzhou (Chica), May 2005, pp. 316–319 (2005)
66. http://www.algolith.com/en/products/intellectual-property
67. http://www.altera.com/literature/ug/ug_vip.pdf
68. http://www.imageproc.com/company.php
69. http://www.zoran.com/IMG/pdf/Frame_It_1.pdf
70. http://www.mstarsemi.com/products/asic_ip
71. http://www.CE.org

72. Yang, S., Lu, T.: SOC design flow case study: design a video processing pipeline. In: Tutorial of International Conference on ASIC (ASICON), Guilin (China), October 2007, p. 10 (2007)
73. Global DTV IC Industry Report, 2007-2008
74. Krishnapuram, R., Keller, J.M.: Fuzzy set theoretic approach to computer vision: an overview. In: Proc. IEEE Int. Conf. on Fuzzy Systems, San Diego, USA, March 1992, pp. 135–142 (1992)
75. Russo, F.: FIRE operators for image processing. Fuzzy Sets and Systems 103, 265–275 (1999)
76. Russo, F.: Edge detection in noisy images using fuzzy reasoning. IEEE Trans. on Instrumentation and Measurement 47(5), 1102–1105 (1995)
77. Russo, F., Ramponi, G.: Edge detction by FIRE operators. In: Proc. IEEE Int. Conf. on Fuzzy Systems, Orlando (USA), June 1994, vol. 26-29, pp. 249–253 (1994)
78. Zadeh, L.A.: Fuzzy logic = Computing with words. IEEE Trans. on Fuzzy Systems 4(2), 103–111 (1996)
79. Tizhoosh, H.R.: Fuzzy image enhancement: an overview. In: Fuzzy techniques in image processing. Studies in Fuzziness and Soft Computing. Physica-Verlag Editorial, Heidelberg (2000)
80. Nachtegael, M., Van der Weken, D., Van De Ville, D., Kerre, E., Philips, W., Lemahieu, I.: An overview of fuzzy filters for noise reduction. In: Proc. IEEE Int. Conf. on Fuzzy Systems, Melbourne, Australia, pp. 7-10 (December 2001)
81. Russo, F.: Fuzzy systems in instrumentation: fuzzy signal processing. IEEE Trans. on Instrumentation and Measurement 45(2) (April 1996)
82. Van De Ville, D., Nachtegael, M., Van der Weken, D., Kerre, E.E., Philips, W., Lemahieu, I.: Noise reduction by fuzzy image filtering. IEEE Trans. on Fuzzy Systems 11(4), 429–436 (2003)
83. Russo, F., Ramponi, G.: A fuzzy operator for the enhancement of blurred and noisy images. IEEE Trans. on Image Processing 4(8) (August 1995)
84. Lee, C.-S., Guo, S.-M., Hsu, C.-Y.: Genetic-based fuzzy image filter and its application to image processing. IEEE Trans. on Systems, Man, and Cybernetics-Part B: Cybernetics 35(4), 694–711 (2005)
85. Arakawa, K.: Fuzzy rule-based image processing. International Journal of Imaging Systems and Technology 8(5), 457–461 (1997)
86. Russo, F.: Recent advances in fuzzy techniques for image enhancement. IEEE Instrumentation and Measurement Magazine 47(6), 1428–1434 (1998)
87. Lee, C.S., Kuo, Y.-H., Yu, P.-T.: Weighted fuzzy mean filters for image processing. Fuzzy Sets and Systems 89, 157–180 (1997)
88. Kuo, Y.H., Lee, C.-S., Chen, C.-L.: High-stability AWFM filter for signal restoration and its hardware design. Fuzzy Sets and Systems 114(2), 185–202 (2000)
89. Lee, C.S., Hsu, C.Y., Kuo, Y.H.: Intelligent fuzzy image filter for impulse noise removal. In: Proc. Int. Conf. on Fuzzy Systems, May 2002, vol. 1, pp. 431–436 (2002)
90. Wang, J.H., Liu, W.J., Lin, L.D.: Histogram-based fuzzy filter for image restoration. IEEE Trans. on Systems, Man and Cybernetics 32(2), 230–238 (2002)
91. Russo, F., Ramponi, G.: A fuzzy filter for images corrupted by impulse noise. IEEE Signal Processing Letter 3(6), 168–170 (1996)
92. Russo, F., Ramponi, G.: Removal of impulse noise using a FIRE filter. In: Porc. IEEE Int. Conf. on Image Processing, Lausanne (Switzerland), September 1996, vol. 2, pp. 975–978 (1996)
93. Mancuso, M., De Luca, R., Poluzzi, R., Rizzotto, G.G.: A fuzzy decision directed filter for impulse noise reduction. Fuzzy Sets and Systems 77, 111–116 (1996)

94. Russo, F.: Hybrid neuro-fuzzy filter for impulse noise removal. Pattern Recognition 32, 1843–1855 (1999)
95. Russo, F.: Noise removal from image data using recursive neurofuzzy filters. IEEE Trans. on Instrumentation and Measurement 49(2), 307–314 (2000)
96. Russo, F.: An image enhancement technique combining sharpening and noise reduction. IEEE Trans. on Instrumentation and Measurement 51(4), 824–828 (2002)
97. Yüksel, M.E., Besdok, E.: A simple neuro-fuzzy impulse detector for efficient blur reduction of impulse noise removal operators for digital images. IEEE Trans. on Fuzzy Systems 12(6), 854–865 (2004)
98. Yüksel, M.E.: A hybrid neuro-fuzzy filter for edge preserving restoration of images corrupted by impulse noise. IEEE Trans. on Image Processing 15(4), 928–936 (2006)
99. Arakawa, K.: Digital signal processing based on fuzzy rules. In: Proc. 5th IFSA World Congress, Seoul (Korea), July 1993, pp. 1305–1308 (1993)
100. Peng, S., Lucke, L.: Multi-level adaptive fuzzy filter for mixed noise removal. In: Proc. Int. Symposium on Circuits and Systems, vol. 2, May 1995, pp. 1524–1527 (1995)
101. Farbiz, F., Menhaj, M.B., Motamedi, S.A.: Edge preserving image filtering based on fuzzy logic. In: Proc. 6th European Congress on Intelligent Techniques and Soft Computing (EUFIT), Aachen (Germany), September 1998, pp. 1417–1421 (1998)
102. Farbiz, F., Menhaj, M.B.: A fuzzy logic control based approach for image filtering. In: Fuzzy techniques in image processing. Studies in Fuzziness and Soft Computing. Physica-Verlag Editorial, Heidelberg (2000)
103. Buckley, J.J.: Universal fuzzy controllers. Automatica 28, 1245–1248 (1992)
104. Ying, H.: Sufficient conditions on general fuzzy systems as function approximators. Automatica 30, 521–525 (1994)
105. Zeng, X.-J., Singh, M.G.: Approximation accuracy analysis of fuzzy systems as function approximators. IEEE Trans. on Fuzzy Systems 4(1), 44–63 (1996)
106. Forsyth, D.A., Ponce, J.: Computer vison: A modern approach. Prentice Hall Editorial, New Jersey (2003)
107. Kerre, E.E., Nachtegael, M. (eds.): Fuzzy techniques in image processing. Studies in Fuzziness and Soft Computing. Physica-Verlag Editorial, Heidelberg (2000)
108. Tyan, C.-Y., Wang, P.P.: Image processing - enhancement, filtering and edge detection using the fuzzy logic approach. In: Proc. IEEE Int. Conf. on Fuzzy Systems, San Francisco (USA), March 1993, vol. 1, pp. 600–605 (1993)
109. Pal, S.K.: Fuzzy sets in image processing and recognition. In: Proc. IEEE Int. Conf. on Fuzzy Systems, San Diego, USA, March 1992, pp. 119–126 (1992)
110. Ting, H.C., Hang, H.M.: Spatially adaptive interpolation of digital images using fuzzy inference. In: Proc. SPIE, Orlando (USA), March 1996, vol. 2727, pt.3, pp. 1206–1217 (1996)
111. Shezaf, N., Abromov-Segal, H., Sutskoner, I., Bar-Sella, R.: Adaptive low complexity algorithm for image zooming at fractional scaling ratio. In: Proc. 21st IEEE Convention of the Electrical and Electronics Engineers, Tel Aviv (Israel), April 2000, pp. 253–256 (2000)
112. Aso, T., Suetake, N., Yamakawa, T.: A code-reduction technique for an image enlargement by using a som-based fuzzy interpolation. In: Proc. 9th Int. Conf. on Neural Information Processing (ICONIP), Torino (Italy), August 1989, vol. 3, pp. 711–721 (1989)
113. Mancuso, M., D'Alto, V., Poluzzi, R.: Fuzzy edge-oriented motion-adaptive noise reduction and scanning rate conversion. In: Proc. IEEE Asia-Pacific Conference on Circuits and Systems (APCCAS), December 1994, pp. 652–656 (1994)
114. Michaud, F., Le Dinh, C.T., Lachiver, G.: Fuzzy detection of edge-direction for video line doubling. IEEE Trans. on Circuits and Systems for Video Technology 7(3), 539–542 (1997)

115. Van De Ville, D., Philips, W., Lemahieu, I.: Fuzzy-based motion detection and its application to de-interlacing. In: Fuzzy techniques in image processing. Studies in Fuzziness and Soft Computing. Physica-Verlag Editorial, Heidelberg (2000)
116. Zadeh, L.A.: Fuzzy sets. Information and Control 8, 338–353 (1965)
117. Lee, C.C.: Fuzzy logic in control systems: fuzzy logic controller. IEEE Trans. on Systems, Man and Cybernetics 20(2), 404–432 (1990)
118. Driankov, D., Hellendoorn, H., Reinfrank, M.: An introduction to fuzzy control. Springer, Heidelberg (1993)
119. Zadeh, L.A.: Outline of a new approach to the analysis of complex systems and decision processes. IEEE Trans. on Systems, Man and Cybernetics 3(1), 28–44 (1973)
120. Cox, E.: The fuzzy systems handbook. Academic Press Profesional, London (1999)
121. Hellendorn, H., Thomas, C.: Defuzzification in fuzzy controllers. Journal of Intelligent and Fuzzy Systems 1, 109–123 (1993)
122. Baturone, I., Sánchez-Solano, S., Barriga, A., Huertas, J.L.: mplementation of inference/defuzzification methods via continuos-time analog circuits. In: Proc. 6th IFSA World Congress, Sao Paulo (Brazil), pp. 623–626 (1995)
123. Takagi, T., Sugeno, M.: Derivation of fuzzy control rules for human operator's control actions. In: Proc. IFAC Symposium on Fuzzy Information, Knowledge Representation and Decision Analysis, Marseille (France), pp. 55–60 (1983)
124. `http://www.fuzzytech.com`
125. `http://www.mathworks.com/products/fuzzylogic`
126. Russo, F.: A user-friendly research tool for image processing with fuzzy rules. In: Proc. IEEE Int. Conf. on Fuzzy Systems, San Diego (USA), March 1992, pp. 561–568 (1992)
127. `http://www.imse.cnm.es/Xfuzzy`
128. Moreno-Velo, F.J., Baturone, I., Sánchez-Solano, S., Barriga, A.: Rapid design of fuzzy systems with Xfuzzy. In: Proc. 12th IEEE International Conference on Fuzzy Systems, St. Louis (USA), vol. 1, pp. 342–347 (2003)
129. Baturone, I., Moreno-Velo, F.J., Sánchez-Solano, S., Barriga, A., Brox, P., Gersnoviez, A., Brox, M.: Using Xfuzzy environment for the whole design of fuzzy systems. In: Proc. IEEE International Conference on Fuzzy Systems (FUZZ-IEEE 2007), London (United Kingdom), pp. 1–6 (2007)

Chapter 2
Fuzzy Motion-Adaptive Algorithm for Video De-Interlacing

Abstract. The ability of fuzzy logic-based systems for video de-interlacing is explored in this Chapter. Particularly, our study is focused on how motion-adaptive de-interlacing can be performed by a fuzzy logic-based system. This Chapter is structured as follows. Firstly, several motion-adaptive strategies are described in Section 2.1. The starting point of our study is the algorithm developed by Van de Ville et al. in [6] described in Section 2.2. It performs a smart de-interlacing by combining a spatial and a temporal interpolator according to the level of motion measured in each pixel of the image. The combination of both interpolators is determined by a fuzzy logic-based system. In Section 2.3, our algorithm based on a fuzzy logic system for motion-adaptive de-interlacing is presented. Its performance is evaluated in Section 2.4 by de-interlacing sequences from video and film material. Finally, some conclusions are expounded in Section 2.5.

2.1 Motion-Adaptive De-Interlacing

As mentioned in the previous Chapter, explicit motion-adaptive algorithms calculate a new pixel value by interpolating between a spatial and a temporal de-interlacing method. For color transmission standards, the motion-adaptive de-interlacing process may be performed separately on each color space component. However, based on the fact that the details of an image are mainly determined by the luminance component of the video signal, and much less by the chrominance components, the motion detection is usually done with the luminance component. Furthermore, this is more efficient in terms of the number of computations required to implement it since only one of the three components is processed. When a de-interlacing method is selected for the luminance component of a pixel, the two chroma components of this pixel are also created by using the same de-interlacing method.

The primitive proposals based on the motion-adaptive idea de-interlace video by applying:

$$I_P = (1-\alpha)I_T + \alpha I_S \tag{2.1}$$

where I_S is the output of a spatial interpolator, and I_T is the output of a temporal interpolator. The variable α, which is the output of a motion detector, ranges from 0 to

P. Brox et al.: Fuzzy Logic-Based Algorithms for Video De-Interlacing, STUDFUZZ 246, pp. 49–84.
springerlink.com

1 and determines the level of motion. The performance of explicit motion-adaptive algorithms relies on the quality of the motion detector, since it is strongly dependent on the combination of both de-interlacing algorithms. The video sequence called 'Carphone', which is usually employed as a benchmark sequence in video processing, is used to show this up. Figure 2.1 illustrates the results obtained for different values of α when the field insertion method is used as I_T and line doubling is used as I_S. The performance of these primitive proposals is very poor since it is highly sensitive to the value of α.

The detection of motion can be improved by processing video on a pixel by pixel basis. This seems logical since motion is normally present in an area of the image while other areas are static. The detection of motion is based on the assumption that motion produces variations in the luminance of pixels with the same spatial co-ordinates but in different fields. Therefore, the simplest motion detector only evaluates the difference between two fields of the same polarity as follows:

$$H(x,y,t) = \frac{|I(x,y,t+1) - I(x,y,t-1)|}{2} \tag{2.2}$$

The function $H(x,y,t)$ involves the pixels shown in Figure 2.2.

One of the initial motion-adaptive de-interlacing algorithms based on a pixel by pixel processing, presented in [1], calculates the new interpolated pixel value as follows:

$$I_p(x,y,t) = (1-\alpha(x,y,t))I_T(x,y,t) + \alpha(x,y,t)I_S(x,y,t) \tag{2.3}$$

where $\alpha(x,y,t)$ is the output of the motion detector for a discrete pixel of coordinates (x,y,t).

A hard switch between the two modes, as shown in Figure 2.3, has a critical effect on the operation of the algorithm. This provides non-effective results, as illustrated in Figure 2.4, where several de-interlaced images of the Carphone sequence have been obtained after applying different threshold values. Again, *field insertion* and *line average* were used as temporal and spatial interpolators.

Several proposals have been presented in the literature to improve the performance of the initial approach with an abrupt transition [1]-[4]. In [1], a continuos motion signal rather than a hard switch is used. It measures motion magnitude by considering the frame differences over at least three adjacent samples (and lines) around the current pixel position and introduces a recursive filter to make sure that when movement starts, the motion detector reacts instantly and decays over several frames when movement ceases. Another solution is to introduce more threshold values are introduced to reduce the effects of using an abrupt transition in [2]. This kind of approaches distinguishes between low and high motion for each pixel in the image (see Figure 2.5(a)). The idea is to take advantage of dealing with intermediate cases of motion from just below to just above the primitive threshold value shown in Figure 2.3. The transition can be softer by using several step functions as shown in Figure 2.5(b) [2]. The number of steps used in the transition can be adjusted accordingly to the transition smoothness desired.

Fig. 2.1. Influence of the motion detector output (α) in a global motion-adaptive de-interlacing.

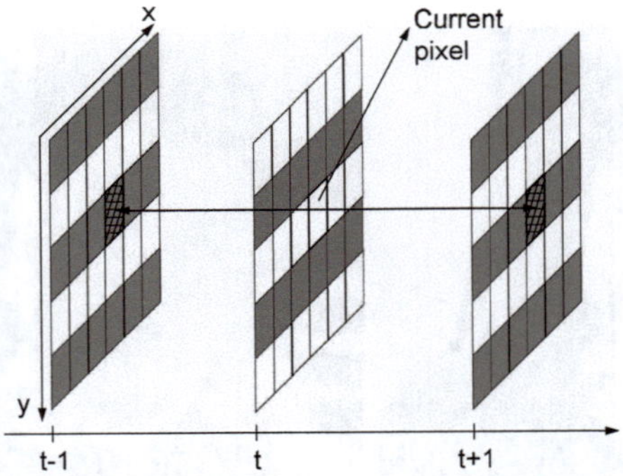

Fig. 2.2. $H(x,y,t)$ measures the difference between the luminance values of two pixels with the same spatial co-ordinates but belonging to different fields.

Another approach is to apply filtering methods, such as the median filter, in the intermediate area between δ_{LOW} and δ_{HIGH} [2]. In this case, the new interpolated pixel is calculated as follows (see Figure 2.6):

$$\begin{aligned} &if(H(x,y,t) \leq \delta_{LOW}) \Rightarrow \alpha(x,y,t) = 0 \Rightarrow I_p(x,y,t) = I_T(x,y,t) \\ &elseif(H(x,y,t) > \delta_{HIGH}) \Rightarrow \alpha(x,y,t) = 1 \Rightarrow I_p(x,y,t) = I_S(x,y,t) \\ &else \Rightarrow I_p(x,y,t) = med(I(x,y-1,t), I(x,y+1,t), I(x,y,t-1)) \end{aligned} \tag{2.4}$$

where *med()* represents the median operator.

Another approach to improve the initial idea presented in [1] has been reported in [3]. It uses the output value of a counter to decide which one of the two modes (I_S or I_T) is used to interpolate. This method implicitly requires for its implementation two threshold values: one to decide if the counter has to be cleared and a second one to decide if the counter value is high enough to use one mode or the other. The

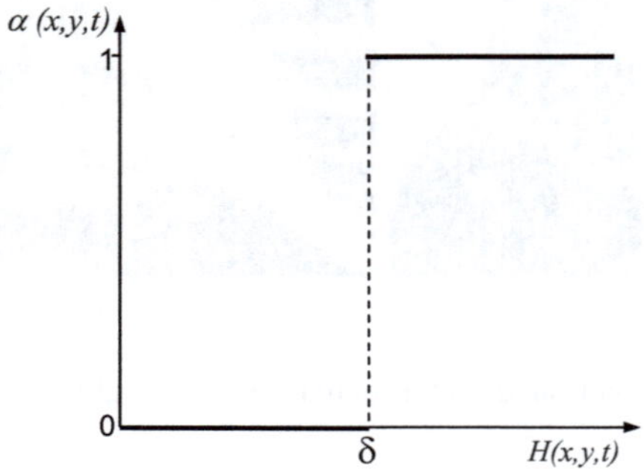

Fig. 2.3. A crisp transition between the two de-interlacing methods.

Fig. 2.4. Influence of the threshold value (δ) in a crisp pixel-by-pixel motion-adaptive de-interlacing.

comparison with both threshold values is made for each pixel in the image, thereby increasing the interpolation capability of the algorithm.

Another approach reported in [4], is to employ adaptive threshold values. This technique performs a global estimation of motion among the reference video fields on a pixel by pixel basis. The result of this estimation decides certain value of the threshold.

Other problems can complicate the suitable behavior of a motion detector such as the presence of noise. To increase the robustness of the motion detectors several proposals have been presented in the literature:

- Different detectors that can work with more than two fields in the neighborhood of the current field [5] can be used. A logical or linear combination of several output detectors provide a more reliable indication of motion.

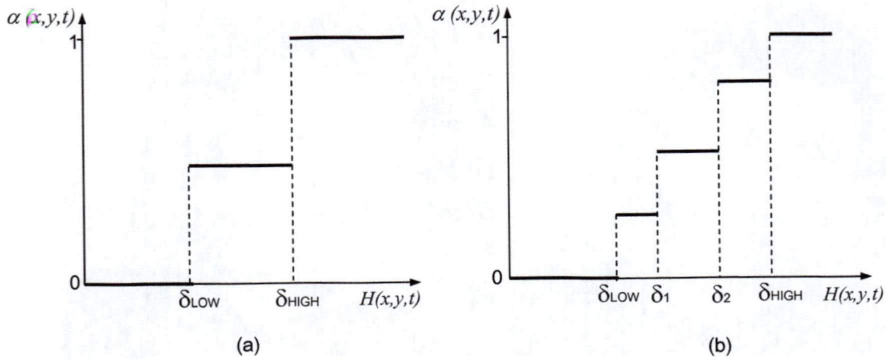

Fig. 2.5. (a) A transition with two threshold values. (b) A transition between the low and high threshold values by using step functions.

- Filtering techniques are used in [5], [2] to filter the input of the motion detector in order to reduce noise.
- The combination of a high and a low pass filter for separately processing spatial and temporal resolution components of the video signal. It is based on the assumption that human vision system tends to loose spatial information for a moving object and pick up spatial details for a static object [2].

A recent contribution to improve the performance of the motion detector was presented in [6],[7] by Van de Ville et al. They propose the use of a fuzzy motion detector inspired by the behavior of a fuzzy logic controller. This motion detector employs different heuristic rules, which combine the information available in the neighborhood of the current pixel to calculate the value of α. The next section explains in detail the Van de Ville et al. proposal, which has been the starting point of our study.

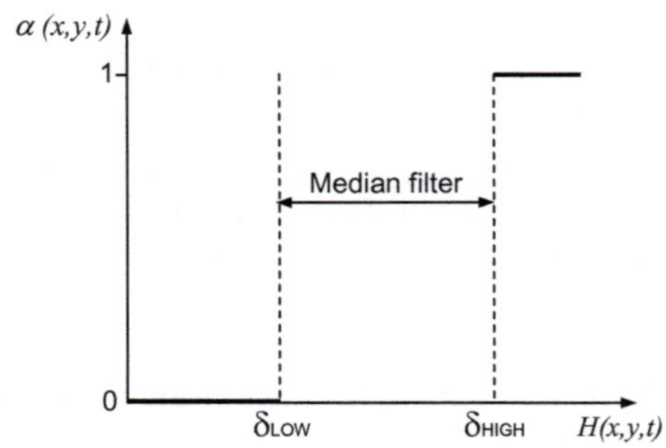

Fig. 2.6. A median filter transition between the two modes of de-interlacing.

2.2 Van de Ville et al. Proposal

Van de Ville et al. propose a fuzzy logic-based system with multiple inputs and a single output (MISO) to evaluate the presence of motion with robustness. More precisely, the system uses $6K+3$ frame difference signals for a new pixel value, where K is the number of neighbors considered at each side of the current pixel.

The first stage of the Van de Ville et al. proposal is the fuzzification process, in which the membership values of the frame difference signals to two fuzzy sets associated to the linguistic labels *SMALL* and *LARGE* are calculated. The membership functions of these fuzzy sets are piece-wise linear, as shown in Figure 2.7. This shape is chosen since it is easily implemented in software or hardware.

To determine the presence of motion, the following knowledge base is applied:

1. When all frame differences in the neighborhood are *SMALL*, the consequent asserts there is no motion present.
2. When all frame differences are *SMALL* at right side of the current pixel but one or more differences at left side are *LARGE*, the consequent states there is no motion.
3. When all frame differences are *SMALL* at left side of the current pixel but one or more differences at right side are *LARGE*, the consequent states there is no motion.
4. When one or more frame differences are *LARGE* at both sides of the current pixel then there is motion present.
5. When the frame difference is *LARGE* at the current pixel, the consequent also assumes the presence of motion.

This rule base for a number K of neighbors at each side of the pixel is summarized in Table 2.1. As it can be seen, the consequents of the rules are symbolized with $M(x,y,t) = true$ (motion is present) or $M(x,y,t) = false$ (motion is absent). For instance, if the number of neighbors is two, the fuzzy rule base is as shown in Table 2.2. In this example, the conjunction ($\wedge$) and disjunction ($\vee$) are represented by the operators AND and OR, respectively.

The output of the fuzzy system proposed by Van de Ville et al. not only computes the consequent for the current pixel $M(x,y,t)$, but also a delayed output $M'(x,y,t)$ stored in an intermediate memory is considered. $M'(x,y,t)$ is calculated as follows:

$$if(M(x,y-1,t-1)\ is\ f) \vee (M(x,y+1,t-1)\ is\ f)\ then\ (M'(x,y,t) = f) \quad (2.5)$$

$$if(M(x,y-1,t-1)\ is\ t) \vee (M(x,y+1,t-1)\ is\ t)\ then\ (M'(x,y,t) = t) \quad (2.6)$$

where 'f' means '$false$' and 't' means '$true$'.

The 'crisp' value of the factor α for the current pixel is finally obtained by using the Fuzzy Mean (FM) defuzzification method as follows:

$$\alpha(x,y,t) = \frac{\pi_{M(x,y,t)}(true)}{\pi_{M(x,y,t)}(true) + \pi_{M(x,y,t)}(false)} \quad (2.7)$$

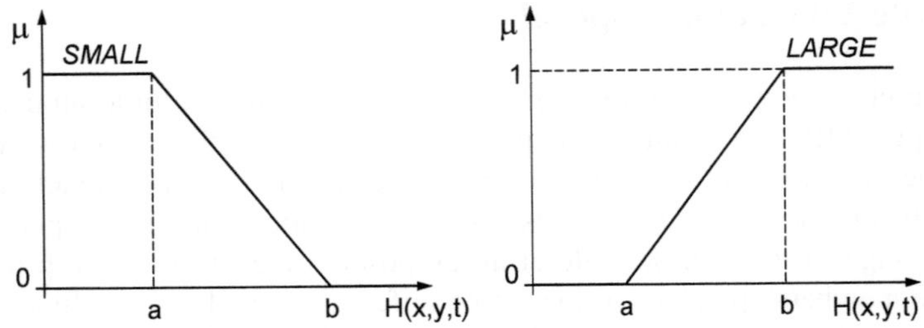

Fig. 2.7. Membership functions for the *SMALL* and *LARGE* fuzzy conceptsdzfgsdfgd.

where $\pi_{M(x,y,t)}(true)$, defined as the combined plausibility of the motion, is calculated as the truth degree of the following fuzzy proposition:

$$(M(x,y,t) \text{ } is \text{ } true) \vee (M'(x,y,t) \text{ } is \text{ } true)) \tag{2.8}$$

and $\pi_{M(x,y,t)}(false)$ is calculated as the truth degree of the following fuzzy proposition:

$$(M(x,y,t) \text{ } is \text{ } false) \wedge (M'(x,y,t) \text{ } is \text{ } false)) \tag{2.9}$$

A total number of $6K+3$ frame differences are considered to evaluate the expressions of plausibility. For instance, if the number of pixels at both sides of the current

Table 2.1. Fuzzy rule set for *K* neighbors at each side of the current pixel.

Rule	Antecedents	Consequent
1.	$\wedge_{i=-K}^{K} H(x+i,y,t)$ is *SMALL*	$M(x,y,t) = false$
2.	$(\wedge_{i=0}^{K} H(x+i,y,t)$ is *SMALL*) $\wedge$ $(\vee_{i=-K}^{-1} H(x+i,y,t)$ is *LARGE*)	$M(x,y,t) = false$
3.	$(\wedge_{i=-K}^{0} H(x+i,y,t)$ is *SMALL*) $\wedge$ $(\vee_{i=1}^{K} H(x+i,y,t)$ is *LARGE*)	$M(x,y,t) = false$
4.	$(\vee_{i=1}^{K} H(x+i,y,t)$ is *LARGE*) $\wedge$ $(\vee_{i=-K}^{-1} H(x+i,y,t)$ is *LARGE*)	$M(x,y,t) = true$
5.	$H(x,y,t)$ is *LARGE*	$M(x,y,t) = true$

Table 2.2. Fuzzy rule set for 2 neighbors at each side of the current pixel

Rule	Antecedents	Consequent
1.	$(H(x+2,y,t)$ is *SMALL*) AND $(H(x+1,y,t)$ is *SMALL*) AND $(H(x,y,t)$ is *SMALL*) AND $(H(x-1,y,t)$ is *SMALL*) AND $(H(x-2,y,t)$ is *SMALL*)	$M(x,y,t) = false$
2.	$(H(x+2,y,t)$ is *SMALL*) AND $(H(x+1,y,t)$ is *SMALL*) AND $(H(x,y,t)$ is *SMALL*)) AND $((H(x-1,y,t)$ is *LARGE*) OR $(H(x-2,y,t)$ is *LARGE*))	$M(x,y,t) = false$
3.	$((H(x+2,y,t)$ is *LARGE*) OR $(H(x+1,y,t)$ is *LARGE*)) AND $(H(x,y,t)$ is *SMALL*) AND $(H(x-1,y,t)$ is *SMALL*) AND $(H(x-2,y,t)$ is *SMALL*)	$M(x,y,t) = false$
4.	$((H(x+2,y,t)$ is *LARGE*) OR $(H(x+1,y,t)$ is *LARGE*)) AND $((H(x-1,y,t)$ is *LARGE*) OR $(H(x-2,y,t)$ is *LARGE*))	$M(x,y,t) = true$
5.	$H(x,y,t)$ is *LARGE*	$M(x,y,t) = true$

pixel is two, the pixels involved in the Van de Ville et al. algorithm are shown in Figure 2.8 (note that 15-frame differences imply 30-pixel luminance values).

Van de Ville et al. show the efficiency of their proposal by de-interlacing several sequences in [7]. After analyzing different values of the parameters K, a, and b (parameters of the membership functions *SMALL* and *LARGE*), they conclude that the best results and, therefore, the lowest errors, are obtained for $K = 2$ and $0 \leq a \leq 1; 6 \leq b \leq 9$.

The proposal of Van de Ville et al. introduces important advantages over other conventional de-interlacing algorithms [7]. However, its weakest point is that implies a high computational complexity since it requires a high number of minimum and maximum operators, as many fuzzification stages as the number of antecedents in the rules, and one division in the defuzzification process to implement the equation (2.7).

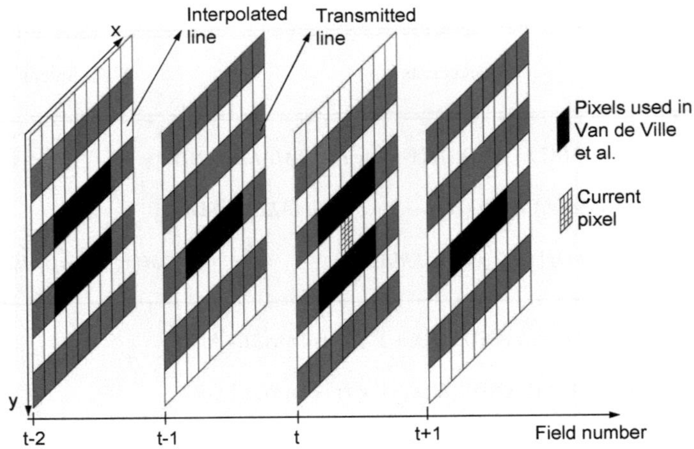

Fig. 2.8. Pixels involved in Van de Ville et al. algorithm for $K = 2$.

2.3 A New Fuzzy Motion-Adaptive De-Interlacing Algorithm

The approach presented herein is inspired by the Van de Ville et al. proposal, but it considerably reduces its computational cost and increases the quality of the motion detector.

Our proposal also considers the influence of neighbors to estimate motion, but it reduces the number of inputs of the system by using bi-dimensional convolution techniques as introduced Gutierrez et al. in [8]. Convolution techniques are very much adequate for the following reasons:

- The use of sum-product instead of min-max operators allows to distinguish better between different levels of motion since the latter only considers extreme values of frame differences.
- Convolution allows to apply different weights to each neighbor when estimating motion. It seems logical to assign higher weights to closer pixels.

Hence, we estimate the level of *motion* as follows:

$$motion(x,y,t) = \frac{\Sigma_{a=1}^{3}(\Sigma_{b=1}^{5} H_{a,b} C_{a,b})}{\Sigma_{a=1}^{3}\Sigma_{b=1}^{5} C_{a,b}} \tag{2.10}$$

where $H_{a,b}$ are the elements of the following frame difference matrix:

$$H = \begin{pmatrix} H_{(-2,-1,-1)} & H_{(-1,-1,-1)} & H_{(0,-1,-1)} & H_{(1,-1,-1)} & H_{(2,-1,-1)} \\ H_{(-2,0,0)} & H_{(-1,0,0)} & H_{(0,0,0)} & H_{(1,0,0)} & H_{(2,0,0)} \\ H_{(-2,1,-1)} & H_{(-1,1,-1)} & H_{(0,1,-1)} & H_{(1,1,-1)} & H_{(2,1,-1)} \end{pmatrix} \tag{2.11}$$

Table 2.3. The proposed fuzzy rule set

Rule	Antecedents	Consequent
1.	$(motion(x,y,t)$ is *SMALL*)	$I_T(x,y,t)$
2.	$(motion(x,y,t)$ is *LARGE*)	$I_S(x,y,t)$

with $H_{(i,j,k)}$ corresponding to the frame difference $H(x+i,y+j,t+k)$ (remember equation (2.2)).

And $C_{a,b}$ are the elements of the following weight matrix:

$$C = \begin{pmatrix} 1\,2\,3\,2\,1 \\ 1\,3\,5\,3\,1 \\ 1\,2\,3\,2\,1 \end{pmatrix} \tag{2.12}$$

The pixels involved in the calculation of this bi-dimensional convolution are the same ones used by Van de Ville et al. (see Figure 2.8). The main advantage is that the use of bi-dimensional convolution reduces the number of input signals of the fuzzy system up to one, so that the fuzzification stage is only realized once per pixel. Furthermore, the proposed rule base is considerably simpler than that presented by Van de Ville et al. (see Table 2.3) since it only contains two rules:

1. If $motion(x,y,t)$ is *SMALL* then the interpolation is realized by applying a temporal de-interlacing algorithm (I_T).
2. If $motion(x,y,t)$ is *LARGE* then the interpolation is realized by applying a spatial de-interlacing algorithm (I_S).

Because of linguistic coherence, both membership functions, *SMALL* and *LARGE*, should be complementary as shown in Figure 2.9. In principle, there is no limitation to

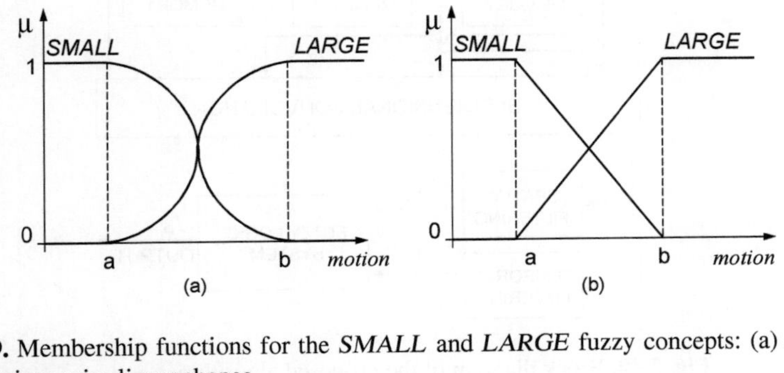

Fig. 2.9. Membership functions for the *SMALL* and *LARGE* fuzzy concepts: (a) S shapes and (b) piece-wise linear shapes.

choose the shape of these functions. Due to its simplicity the piece-wise linear functions in Figure 2.9(b) have been finally selected to describe the fuzzy sets *SMALL* and *LARGE*.

The new luminance component of a de-interlaced pixel provided by the system is calculated by applying the FM defuzzification method (which corresponds to a weighted average), as follows:

$$I_P(x,y,t) = \alpha_1 I_T(x,y,t) + \alpha_2 I_S(x,y,t) \tag{2.13}$$

where α_1 and α_2 are the activation degrees of the first and the second rule, respectively.

If both membership functions are complementary, that is, their superposition is always the unitary value, only one of the rules has to be processed and the expression in (2.13) can be replaced by the following:

$$I_P(x,y,t) = (1-\alpha_2)I_T(x,y,t) + \alpha_2 I_S(x,y,t) \tag{2.14}$$

It is important to emphasize that the expression in (2.14) is equivalent to that proposed in equation (2.3) but now with continuous values of α. That is, the output of the proposed fuzzy system implements a motion-adaptive algorithm in which the output of the motion detector is calculated as the activation degree of the rule 2 in Table 2.3. The shape of '*alpha* (α)' as a function of motion depends on the selected shape of the membership functions used to model the *SMALL* and *LARGE* fuzzy concepts. In case that the chosen membership functions are as shown in Figure 2.9(b), the transition between the value '0' and '1' of α is linear with motion.

A block diagram of the proposed algorithm is shown in Figure 2.10. Let us analyze the complexity of our proposal in comparison with other well-known algorithms. In particular, line doubling and line average have been considered as spatial or intra-field de-interlacing methods. Furthermore, two spatial edge-adaptive methods have been used: the conventional ELA with 3+3 and with 5+5 pixels from the upper and lower lines [9]. Field insertion is considered as the temporal de-interlacing

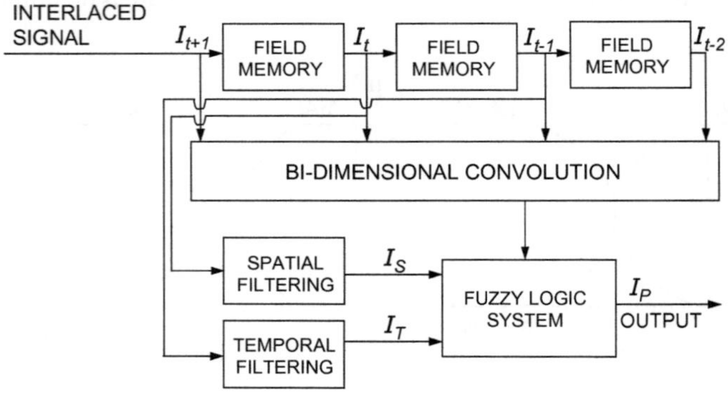

Fig. 2.10. Block diagram of the proposed algorithm.

Table 2.4. Analysis of complexity of different de-interlacing algorithms

	Resources			Primitive Operations (POs)
De-interlacing Algorithm	Line buffer	Field memory	Register	
Line doubling	-	-	-	-
Line average	1	-	-	2
ELA 3+3	1	-	2	11
ELA 5+5	1	-	4	20
Field insertion	-	1	-	-
VT 2 fields	2	1	-	16
VT 3 fields	2	2	-	12
Median motion	1	1	-	11
Van de Ville et al.	1	2+1	4	103
Proposal	1	3	4	45

method. The vertico-temporal algorithms with two [10] and three fields [11], the implicit motion-adaptive algorithm in [12], and the proposal of Van de Ville et al. have also been considered (the latter with the following parameters $K = 2$, $a = 0.5$ and $b = 8$). Our proposal also uses the same values for the parameters (a,b) of the membership functions, *line average* as I_S and *field insertion* as I_T.

Complexity is analyzed using two figures of merit: 1) resources to store pixel values; and 2) primitive operations (POs). Table 2.4 shows the storage devices that are required by each de-interlacing algorithm used in the comparison. As it can be seen, spatial de-interlacing algorithms do not require field memories since they work with pixels from the current field. However, the rest of proposals use information from temporal neighbors and need field memories to store them. The Van de Ville et al. algorithm uses two field memories to evaluate motion at the current pixel, and an intermediate memory to store the output of the fuzzy system in the previous field (see expressions 2.5 and 2.6). Our proposal also requires three field memories to compute the bi-dimensional convolution.

Algorithm complexity can be also estimated by evaluating the number of POs to calculate an interpolated pixel. Addition, subtraction, shifting, absolute difference and sign function have been considered as operations of complexity 1, while multiplication and division have been considered of complexity 2. As can be seen in Table 2.4, our proposal slightly increases the POs of vertico-temporal algorithms, and considerably reduces the number of POs in comparison with the Van de Ville et al. algorithm.

2.4 Simulation Results

2.4.1 Results on Benchmark Video Sequences

The performance of the proposed algorithm has been analyzed by de-interlacing several video sequences that have been widely used as benchmarks in video processing applications. The origin of the test material is categorized into two types. The first group is a set of progressive sequences that have been interlaced in order to measure the difference between the obtained interpolated frames and the original ones. Many error measures have been proposed as figures of merit to evaluate the quality of digital video. However, as a consequence of the complexity of the human visual system, it is difficult to find an objective criterion to entirely quantify image distortions. Nevertheless, some measures seem to have a higher correlation than others with the perceived quality. A very popular quality criteria is the MSE (Mean Squared Error) between the original and the reconstructed images, which is given by the following expression:

$$MSE(t) = \frac{1}{M \cdot N} \Sigma_{x,y} (I_P(x,y,t) - I_{original}(x,y,t))^2 \tag{2.15}$$

where the image has a resolution of MxN pixels, I_P is the value of the interpolated pixel, and $I_{original}$ is the pixel value in the original progressive image. Strongly related to the MSE is the PSNR, which is defined as follows:

$$PSNR(t) = 20 \log \frac{255}{\sqrt{MSE(t)}} \tag{2.16}$$

Actually a unique PSNR value is not meaningful, but the comparison between two values for different reconstructed images gives a measurement of quality. Generally, an improvement of 0.5 dBs in PSNR is quite perceptible by the human visual system in the de-interlaced image. Figure 2.11 illustrates the strategy used to test the proposed algorithm.

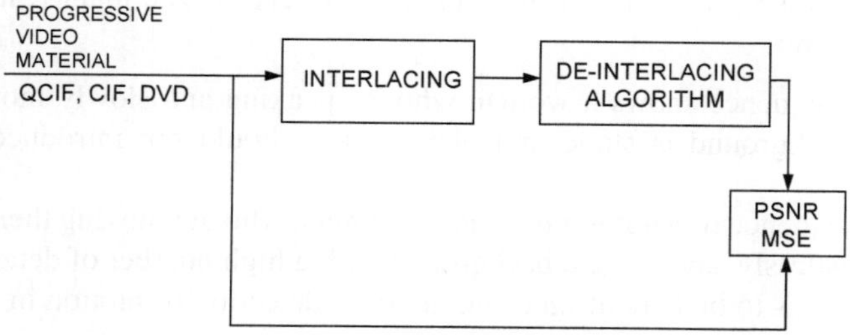

Fig. 2.11. Strategy used to test the proposed algorithm.

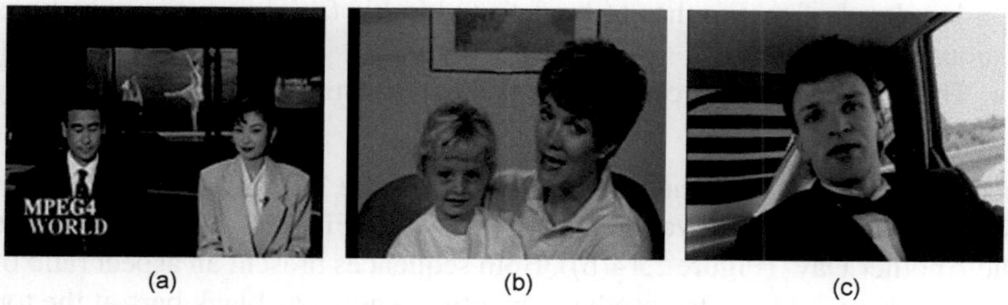

Fig. 2.12. One snapshot of the QCIF sequences: (a) News, (b) Mother, and (c) Carphone.

Table 2.5 shows the average PSNR values obtained when de-interlacing 50 fields of different video sequences with four different formats: QCIF, CIF[1], DVD PAL, and DVD NTSC[2]. DVD equipments provide a progressive scanning, whose format depends on the TV transmission system: DVD PAL (576x720) and DVD NTSC (480x720).

One snapshot of each of the QCIF sequences is shown in Figure 2.12. Next, the main characteristics of these sequences are briefly described:

- News: This sequence shows a typical scene of news on TV. Two newsreaders slowly move their faces whereas the background combines a static area with a screen where a couple of dancers is moving. The fuzzy system has to be able to deal with the details of the background as well as avoid false cases of motion detection in critical areas such as the lettering at the bottom of the frame.
- Mother: This sequence corresponds to a mother with her daughter. The mother slowly moves one of her arms above her daughter's head whereas she turns her head quickly at the end of the sequence. The background is static and does not present many difficulties since it is plain except from the horizontal edges of the painting.
- Carphone: This sequence is trickier than the others to de-interlace. It presents high contrast diagonal edges with a low level of motion in the rear and lateral window, and a high level of motion around the man's face.

[1] QCIF and CIF sequences were taken from http://decsai.ugr.es.
[2] DVD material was provided by Philips Research Eindhoven

Figure 2.13 shows a snapshot of the four CIF sequences. The main features of these sequences are:

- Missa: This sequence shows a woman who is speaking and slowly moves her head. The background is static and plain and it should not introduce many difficulties.
- Paris: This sequence illustrates a man and a woman, who are moving their hands and heads restlessly, and a static background with a high number of details. The fuzzy system has to be able of ignoring the false detection of motion in the stationary area.
- Salesman: This sequence shows a salesman holding a cubic box in his hand. Motion is mainly focused on the arm of the salesman and the cubic box. Furthermore, a low level of motion has to be detected in his facial features since he is continuously speaking.
- Trevor: This sequence also presents a speaker and a static background with many vertical edges.

The sequences with DVD formats are shown in Figure 2.14. They correspond to two very famous movies 'Star Wars: Phantom Menace' (Figure 2.14(a)) and 'James Bond: Die Another Day' (Figure 2.14(b)). Both sequences present an aspect ratio of the picture equals to 2.35:1. This produces an effect known as black bars at the top

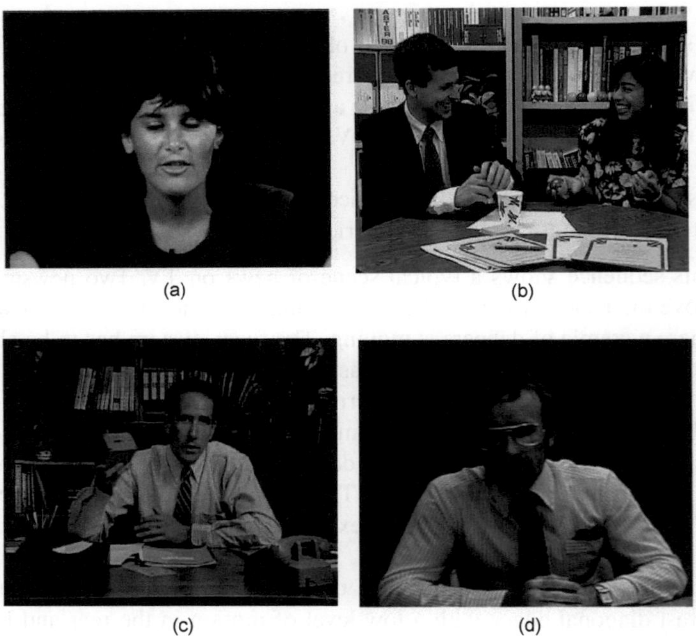

Fig. 2.13. One snapshot of the CIF sequences: (a) Missa, (b) Paris, (c) Salesman, and (d) Trevor.

(a) (b)

Fig. 2.14. One snapshot of the DVD sequences: (a) Phantom Menace and (b) Die Another Day.

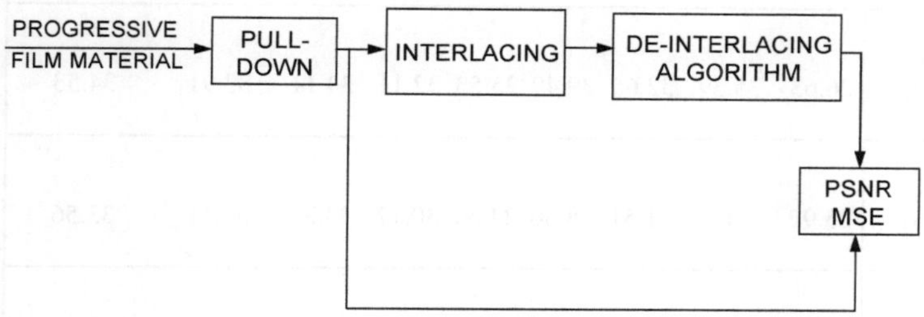

Fig. 2.15. Strategy used to test sequences from film material.

and bottom of the screen when they are displayed on a 16:9 or 4:3 screen as shown in Figure 2.14. Both sequences show horizontally moving scenes with many details.

To carry out the comparison, several de-interlacing algorithms have been employed.

Let us comment the PSNR results shown in Table 2.5. The results show that the proposed algorithm performs better than the other algorithms since it achieves the highest values. Particularly, it considerably improves the results of all the spatial de-interlacing methods in 4 dBs or more, including the edge-adaptive algorithms. Our proposal also exceeds the results achieved by vertico-temporal algorithm with two fields. The vertico-temporal algorithm with three fields only improves slightly our results in two sequences (Missa and Trevor). Our proposal performs better than other algorithms in the rest of the sequences.

The comparison with the median-based method is also favorable. In none of the sequences this approach achieves a higher PSNR value. Finally, our proposal improves the results achieved by Van de Ville et al., introducing an average improvement of almost 2 dBs in the majority of the sequences.

Table 2.5. PSNR results (in dBs) for different de-interlacing methods

	News	Mother	Car-phone	Missa	Paris	Sales-man	Trevor	Phantom Menace	Die AnotherDay
	QCIF	QCIF	QCIF	CIF	CIF	CIF	CIF	DVD PAL	DVD NTSC
Line doubling	25.18	31.81	28.25	36.44	23.61	29.75	31.05	29.57	31.76
Line average	29.25	35.94	32.61	40.47	26.67	35.04	33.53	34.14	35.18
ELA 3+3 [9]	26.63	35.39	32.65	39.49	25.53	32.11	34.11	33.91	34.53
ELA 5+5 [9]	25.92	34.2	31.51	38.56	24.64	30.17	33.31	33.11	33.56
Field insertion	33.13	36.14	30.34	38.36	29.86	36.17	34.36	25.78	39.63
VT 2 fields [10]	35.46	39.61	34.08	40.25	30.73	36.54	36.61	32.11	39.51
VT 3 fields [11]	35.67	40.89	34.54	40.52	31.37	37.16	36.95	34.92	37.82
Median motion [12]	33.51	38.49	33.31	39.44	30.27	36.61	35.43	32.46	41.44
Van de Ville et al. [6]	34.73	39.49	32.27	40.01	33.12	37.62	35.38	34.57	39.52
Proposal	37.51	41.87	34.78	40.18	35.28	38.29	36.69	35.01	42.27

Fig. 2.16. Images from each test sequence: (a) Fire Rose, (b) Fargo1, (c) Fargo2, (d) Tokyo, and (e) Bodyguard.

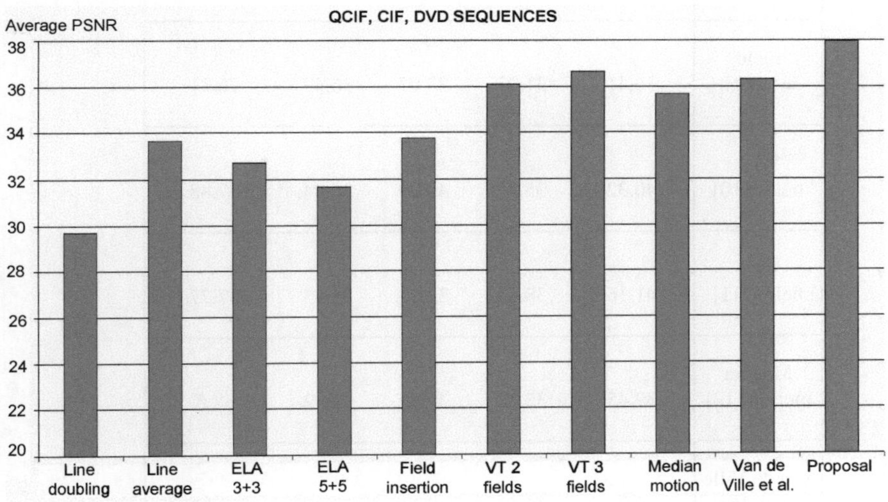

Fig. 2.17. Average PSNR value achieved by each de-interlacing method for QCIF, CIF, and DVD sequences.

A second type of sequences analyzed corresponds to interlaced material that has been directly taped from TV[3]. All these sequences are film material that was converted into PAL system by employing the pull-down 2:2 process. This means that

[3] This material was also provided by Philips Research Laboratories.

Table 2.6. PSNR results (in dBs) for different de-interlacing methods

	Fire Rose	Fargo1	Fargo2	Tokyo	Bodyguard
	PAL TV	PAL TV	PAL TV	PAL TV	PAL TV
Line doubling	34.51	30.48	28.79	27.22	31.45
Line average	38.76	35.92	34.31	31.46	35.61
ELA 3+3 [9]	35.55	35.28	33.66	30.02	35.08
ELA 5+5 [9]	33.61	34.33	32.16	28.53	33.93
Field insertion	36.41	31.23	33.07	36.49	26.31
VT 2 fields [10]	40.32	35.87	40.99	36.84	37.88
VT 3 fields [11]	41.16	38.43	38.91	35.13	37.77
Median motion [12]	39.45	35.32	39.89	34.92	38.5
Van de Ville et al. [6]	39.36	36.64	40.11	34.88	39.05
Proposal	41.14	37.33	42.54	37.71	40.28

the original progressive material can be easily obtained by applying field insertion method with the previous or the next field in the sequence. The strategy used to test our proposal with this kind of sequences is illustrated in Figure 2.15.

A single picture of each sequence is shown in Figure 2.16. The main features of these sequences are briefly the following:

- Fire Rose: This sequence is very complex to de-interlace since the man is moving slowly. Moreover it presents a high number of details in the priest's beard, in which the detection of low level of motion is very tricky.
- Fargo1: It corresponds to a scene of the 'Fargo' movie shooting from inside a car. This scene contains detailed areas in the road signs, and high contrast edges in the steering wheel of the car.
- Fargo2: This sequence also belongs to the 'Fargo' movie. It contains high contrast diagonal lines in the hood of the cars.
- Tokyo: It is an aerial shot from the Tokyo city. The scene moves horizontally and contains many vertical details.
- Bodyguard: It is a scene from the famous movie 'The bodyguard'. This sequence has been selected since it includes subtitles, and it is important to test how the de-interlacing algorithms deal with detailed letterings.

Table 2.6 shows a comparison of the algorithms using the average PSNR when fifty fields of the sequences are de-interlaced.

Again, the results demonstrate a superior performance of our approach over the spatial and temporal methods. Furthermore, it clearly improves the results of the vertico-temporal approaches with 2 fields, the median-based method, and the Van de Ville et al. algorithm. The vertico-temporal with 3 fields performs slightly better in the Fire Rose and Fargo1 sequences, but our approach is significantly better in the rest of the sequences.

To better illustrate the performance of each de-interlacing method, let us consider the bar diagrams shown in Figure 2.17 and 2.18 where each bar corresponds to the

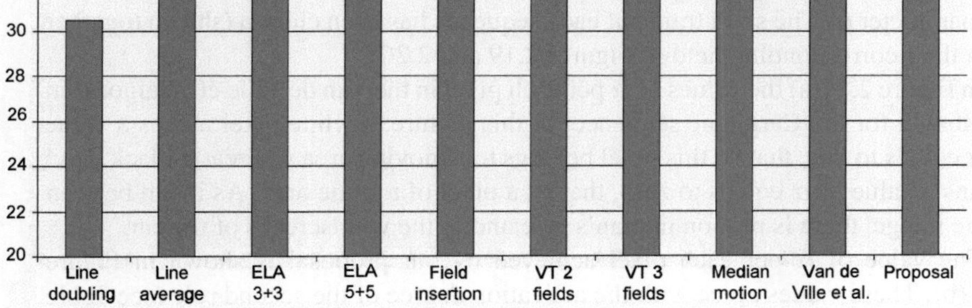

Fig. 2.18. Average PSNR value achieved by each de-interlacing method for film sequences.

Fig. 2.19. Selected frames from Carphone (a) and Salesman (b) sequences and their corresponding fields.

average PSNR value achieved by a de-interlacing technique. The results show that our proposal provides an overall improvement of more than 1 dB for both types of sequences: film and video.

2.4.2 *Detailed Results on Three Sequences*

Let us show the efficiency of our algorithm by detailing the results obtained with three sequences: Carphone, Salesman, and Fargo2.

Firstly, the quality of the motion-adaptive algorithms to detect the moving areas is corroborated by displaying the output of the motion detector, that is, the value of the parameter α. The sixth frame of each sequence has been chosen (shown together with their corresponding field in Figures 2.19 and 2.20).

In Figure 2.21(a) the values of α per each pixel in the Van de Ville et al. algorithm are shown for the Carphone sequence. In this picture, a white pixel means a value of α equals to one, that is, this pixel belongs to a moving area whereas a black pixel means a value of α equals to zero, that is, a pixel of a static area. As it can be seen in the image, there is motion in man's face and in the windscreens of the car.

The value of α per each pixel achieved by our proposal is shown in Figure 2.21(b). These values agree with the activation degree of the second rule (see Table 2.3) of our fuzzy system as it was explained before. From the inspection of both images a main improvement can be identified: our proposal detects more different levels of motion (gray range of colors in Figure 2.21(b)). The highest level of motion is located on the man's face whereas a lower level of motion is detected in the windscreen. Figure 2.21(a) makes evident that the Van de Ville et al. proposal is unable to distinguish among several levels of motion. As a consequence of using min-max operators, it tends to provide the extreme values of motion.

The same analysis has been carried out for the other sequences. Figure 2.22(a) shows the values of α for the field in the Salesman sequence when the Van de Ville et al. algorithm is used. It corroborates the presence of motion in the man's body. However, the α values achieved by our proposal also provide information about

FRAME

FIELD

Fig. 2.20. Selected frame from Fargo2 sequence and its corresponding field.

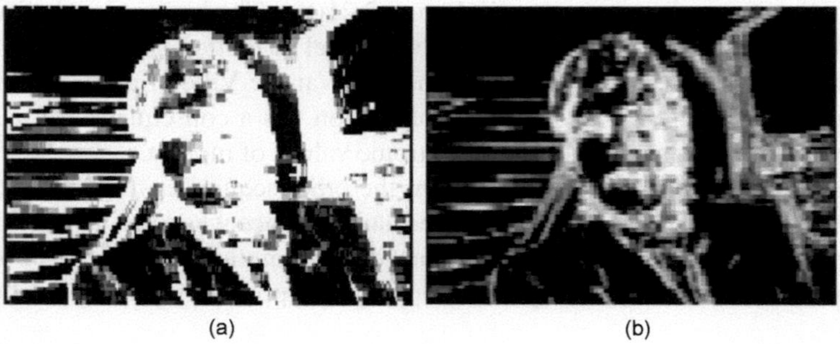

Fig. 2.21. Carphone sequence: (a) Value of α in the Van de Ville et al. algorithm. (b) Value of α achieved by our proposal.

the level of motion as shown in Figure 2.22(b). The high motion areas are in the man's head and in his hands, whereas motion in the shirt is barely perceptible. It is important to emphasize that our fuzzy system is more robust than the proposed by Van de Ville et al., since it considerably reduces false cases of motion due to vertical details in the background.

The detection of motion by the Van de Ville et al. approach is shown in Figure 2.23(a) for the Fargo2 sequence. The contribution of our proposal is shown in Figure 2.23(b). In this case, the main level of motion is located on the men's bodies. Again our approach is more robust to avoid false cases of motion detection in the background.

An analysis of the MSE value evolution has been carried out for these three sequences, Carphone, Salesman, and Fargo2. The graphs in Figure 2.24 show the MSE value for each field of the Carphone sequence. This Figure compares our proposal with the spatial and temporal methods. The MSE results achieved by our approach show up the efficiency of motion-adaptive methods by combining spatial and

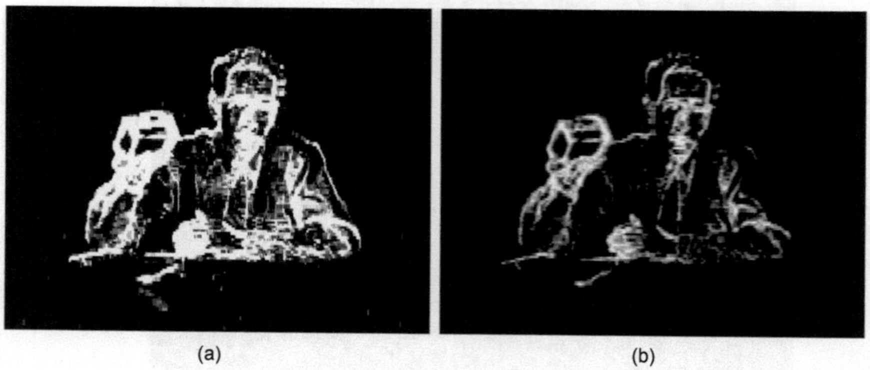

Fig. 2.22. Salesman sequence: (a) Value of α in the Van de Ville et al. algorithm. (b) Value of α achieved by our proposal.

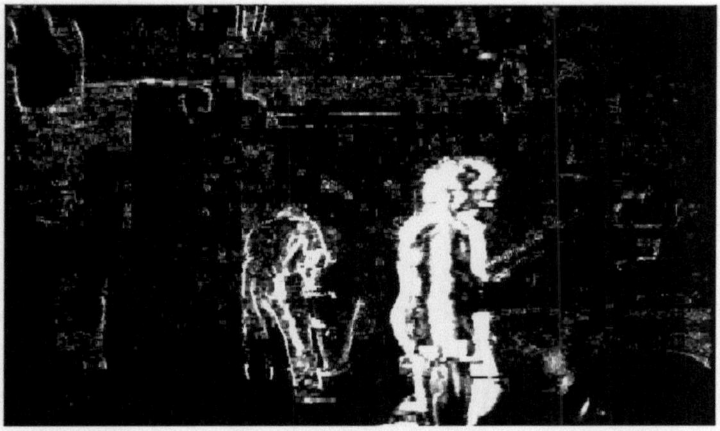

(a)

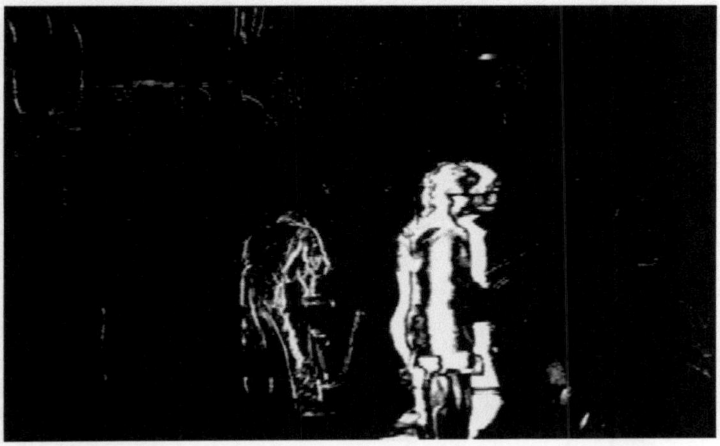

(b)

Fig. 2.23. Fargo2 sequence: (a) Value of α in the Van de Ville et al. algorithm. (b) Value of α achieved by our proposal.

temporal techniques according to the presence of motion. Among spatial algorithms, line doubling presents the worst results, whereas line average offers a more attractive solution and considerably reduces the error values. Edge-adaptive methods do not provide competitive numerical results in comparison with the line average method, but they significantly improve the quality of the reconstructed edges. Let us to corroborate this argument by means of a visual inspection of the de-interlaced pictures in Figure 2.26. Field insertion technique achieves a low error value when the movement between the fields is scarce (for instance the fields with numbers 7 and 24), but the MSE values increase excessively when the level of motion is high (field 43).

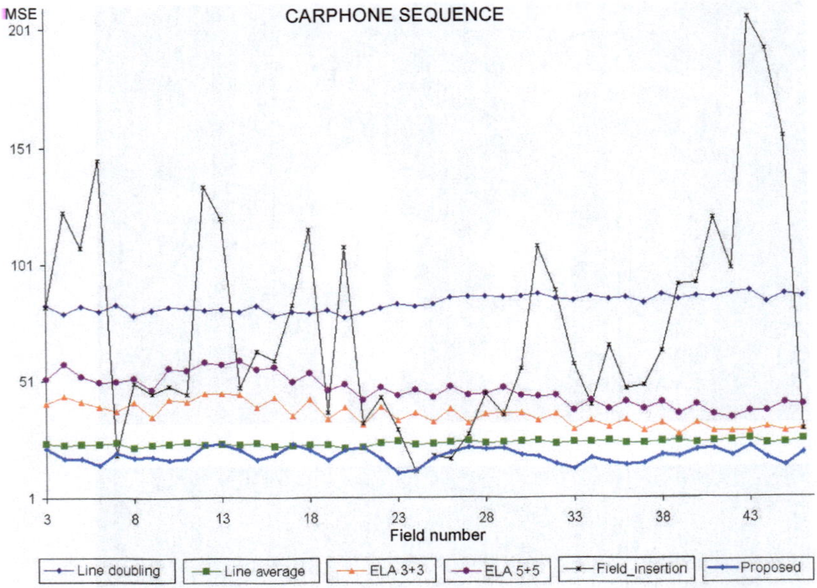

Fig. 2.24. Carphone sequence: comparison with spatial and temporal methods.

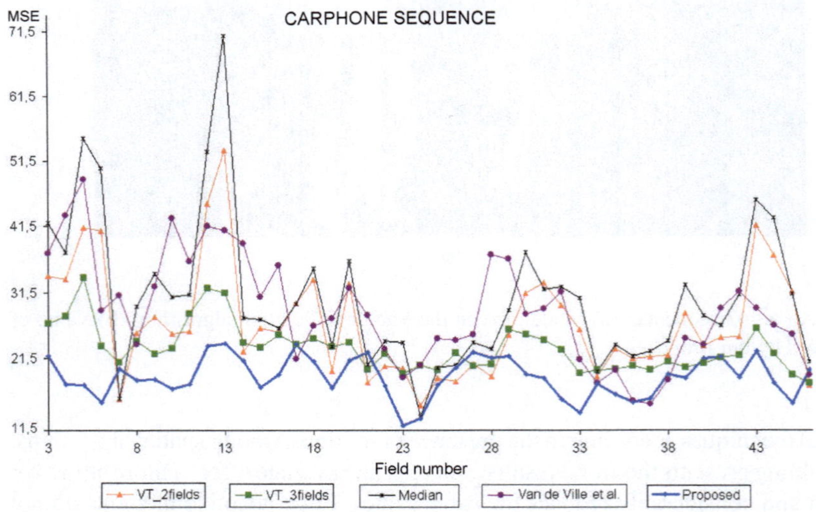

Fig. 2.25. Carphone sequence: MSE results using vertico-temporal and motion-adaptive methods.

The comparison with vertico-temporal and motion-adaptive methods is shown in the graphs of the Figure 2.25. They achieve better results than spatial and temporal methods, but all of them are improved by our proposal. The vertico-temporal method

Fig. 2.26. De-interlaced images obtained with several de-interlacing techniques for the Carphone sequence.

with three fields performs slightly better than the vertico-temporal method with two fields, the median-based technique, and even the Van de Ville et al. proposal.

Let us focus on the visual inspection of the resultant de-interlaced images. Figure 2.26 shows the de-interlaced images obtained by the different algorithms for the Carphone sequence. Line doubling and line average do not reconstruct edges properly. As could be expected, the edge-adaptive algorithms provide a better reconstruction of edges, but they introduce some annoying mistakes such as black points

Fig. 2.27. De-interlaced images for the Carphone sequence.

in the frames of the windows. Field insertion technique produces the typical 'ghosting' effect in moving areas. This artifact is eliminated by vertico-temporal technique with two fields.

Figure 2.27 shows the results obtained by the following de-interlacing methods: the vertico-temporal with three fields, the median-based method, the Van de Ville et al. algorithm, and the proposed algorithm. All these spatio-temporal techniques overcome the results achieved by spatial and temporal methods. Among them, the best results are achieved by VT-3 fields and our proposal, as shown in Figure 2.27.

The graphs in Figure 2.28 show the MSE value for each field of the Salesman sequence. Among spatial methods, the line average method is again the best for this sequence. Even though field insertion achieves good results for the last thirty fields of the sequence where the level of motion is considerably low, the proposed algorithm again provides the best results. It is important to emphasize that the proposed motion-adaptive algorithm is able to combine two bad de-interlacing algorithms to considerably improves their results.

The graphs in Figure 2.29 show the comparison with vertico-temporal and motion-adaptive algorithms. In this case, both motion-adaptive proposals that use fuzzy logic clearly perform better than the rest. Though the previous results could

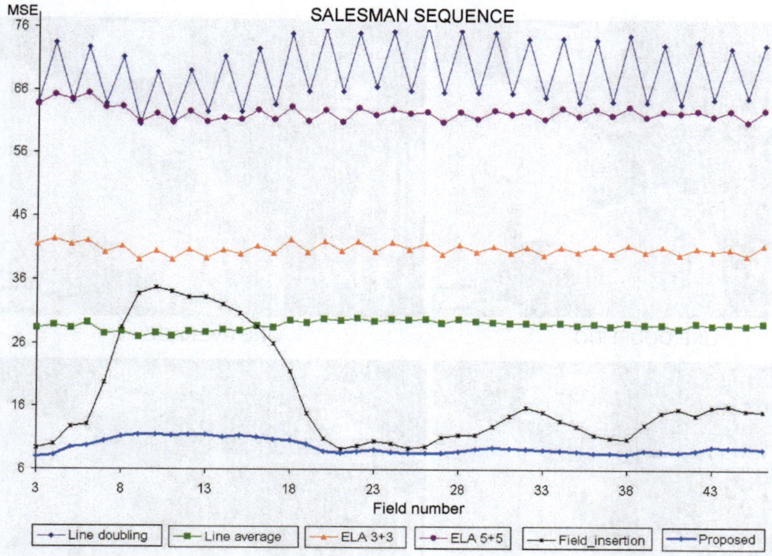

Fig. 2.28. Salesman sequence: comparison with spatial and temporal methods.

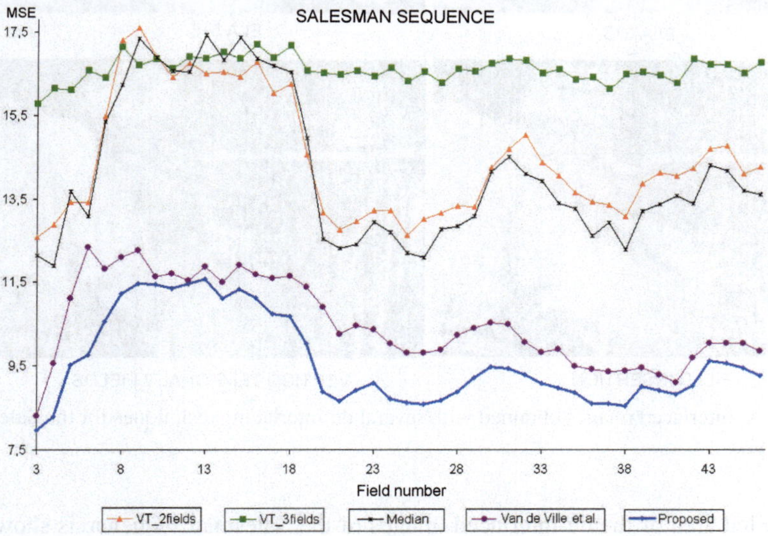

Fig. 2.29. Salesman sequence: MSE results using vertico-temporal and motion-adaptive methods.

lead us to think that vertico-temporal with three fields always offers the best results, it is overcome by the median-based method and vertico-temporal technique with two fields in this particular sequence.

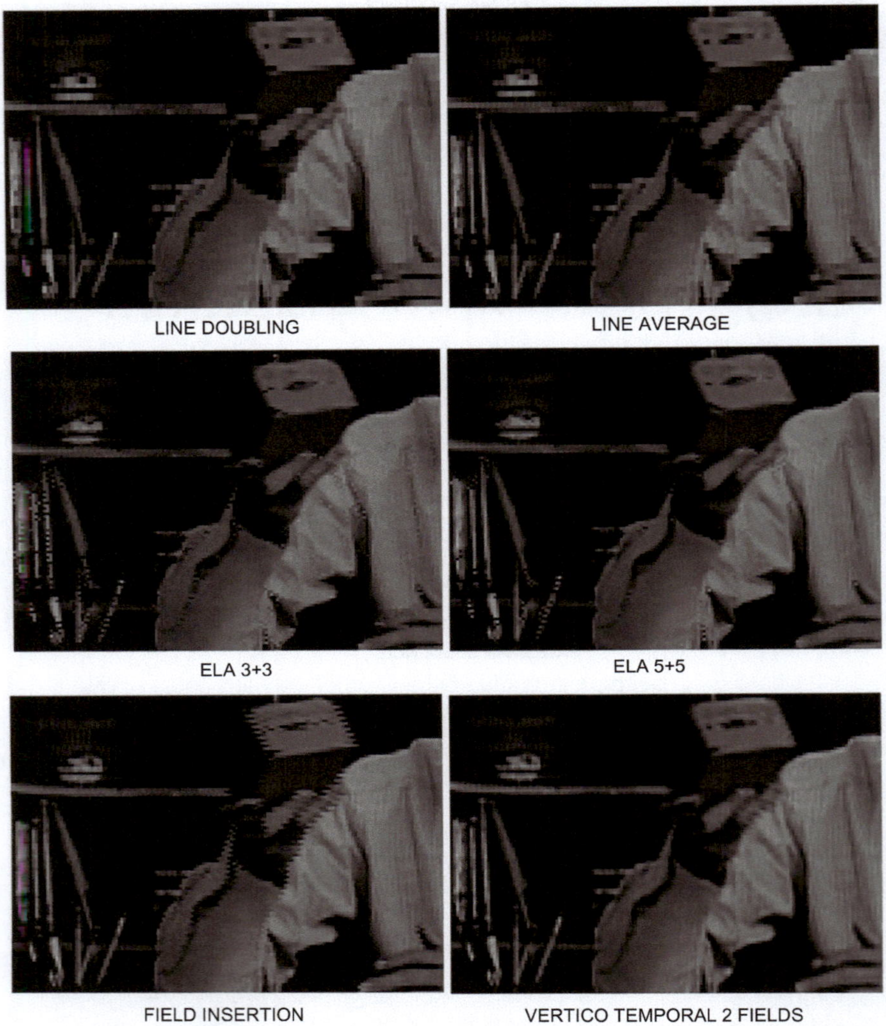

Fig. 2.30. De-interlaced images obtained with several de-interlacing techniques for the Salesman sequence.

A detailed area of the de-interlaced images of the Salesman sequence is shown in Figures 2.30 and 2.31. Line doubling and line average do not provide a suitable reconstruction of the areas with details (for instance, the pencils or books on the shelves). Edge-adaptive methods de-interlace better the clear edges but they fail in the interpolation of non-clear edges (for example, the boundaries of the pencils). Field insertion technique produces the ghosting effect in the moving areas (the box and the salesman's hand). Figure 2.30 also shows the vertico-temporal method with two fields.

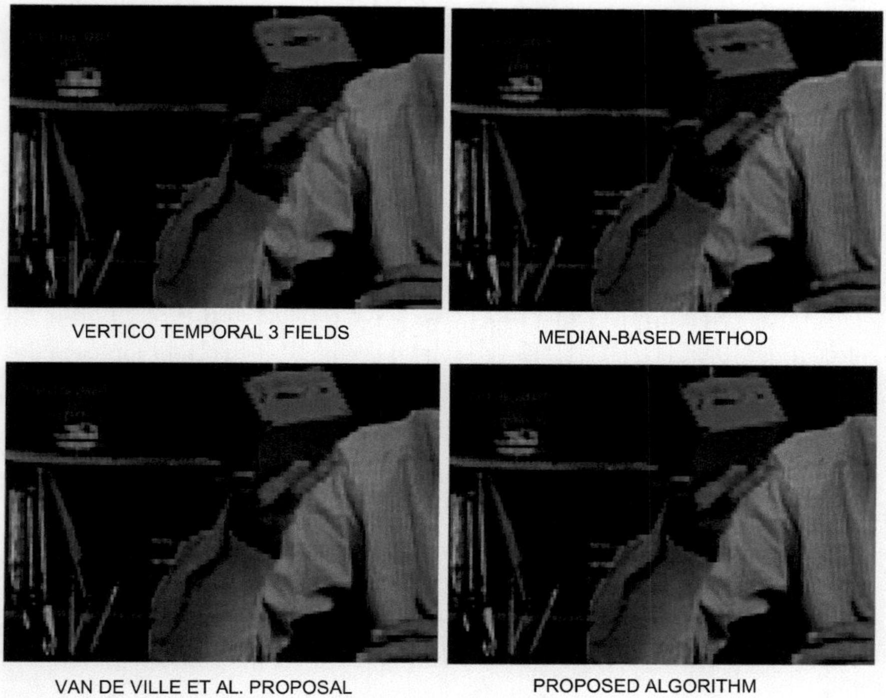

Fig. 2.31. De-interlaced images for the Salesman sequence.

Figure 2.31 shows the vertico-temporal methods with three fields, the median-based method, and Van de Ville et al. proposal. All these proposals overcome the quality of the images obtained with the spatial and temporal algorithms in Figure 2.31, but it is difficult to appreciate a significant improvement among them by visual inspection of the de-interlaced images.

Finally, let us show the results for the Fargo2 film sequence. The graphs in Figure 2.32 show a comparison of the spatial and temporal methods using the MSE criterion in each field of the sequence. Among the spatial methods, line average again presents the best results, but in this case the ELA values are quite close. However, the highlight among these results is the performance of the field insertion technique, which achieves a perfect de-interlaced image for the odd fields but a very high MSE for the even ones. This behavior is a consequence of the repetition of fields in the pull-down 2:2 process.

As shown the graphs in Figure 2.33, the methods that use temporal neighbors pixels from the previous field (VT-2fields, the median based method, the Van de Ville et al. proposal, and our proposal) also present this oscillating behavior. This means that they perform rather well when the previous field is equal to the current one, but rather bad when both fields are different. The vertico-temporal approach

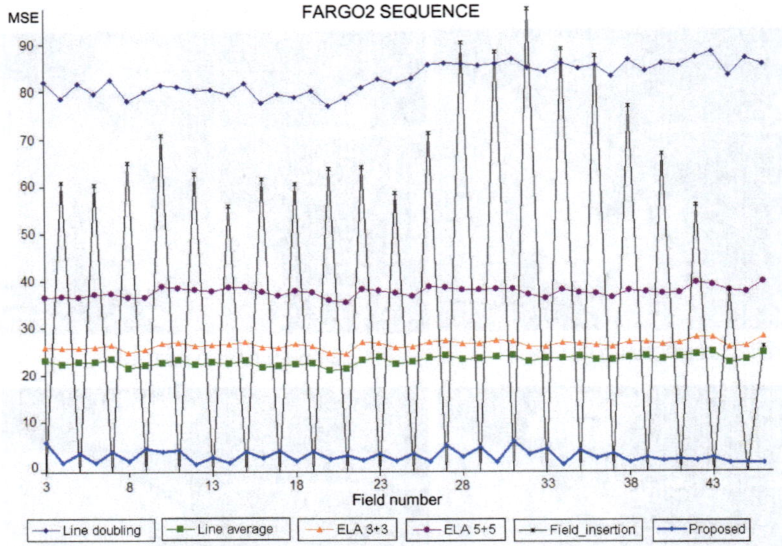

Fig. 2.32. Fargo2 sequence: comparison with spatial and temporal methods.

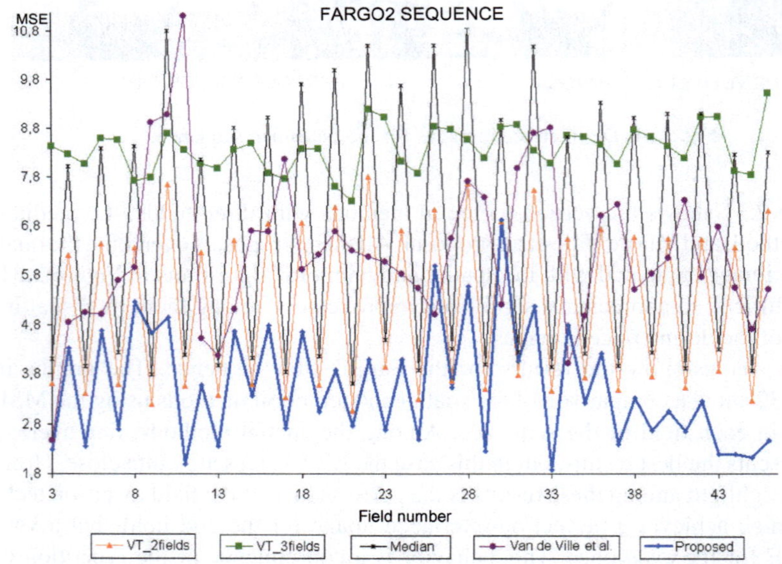

Fig. 2.33. Fargo2 sequence: MSE results using vertico-temporal and motion-adaptive methods.

with three fields does not present such peaks since it works with three fields (two of them similar and the other different).

Although our proposal presents this oscillating effect, it achieves the lowest errors as shown in Figure 2.33.

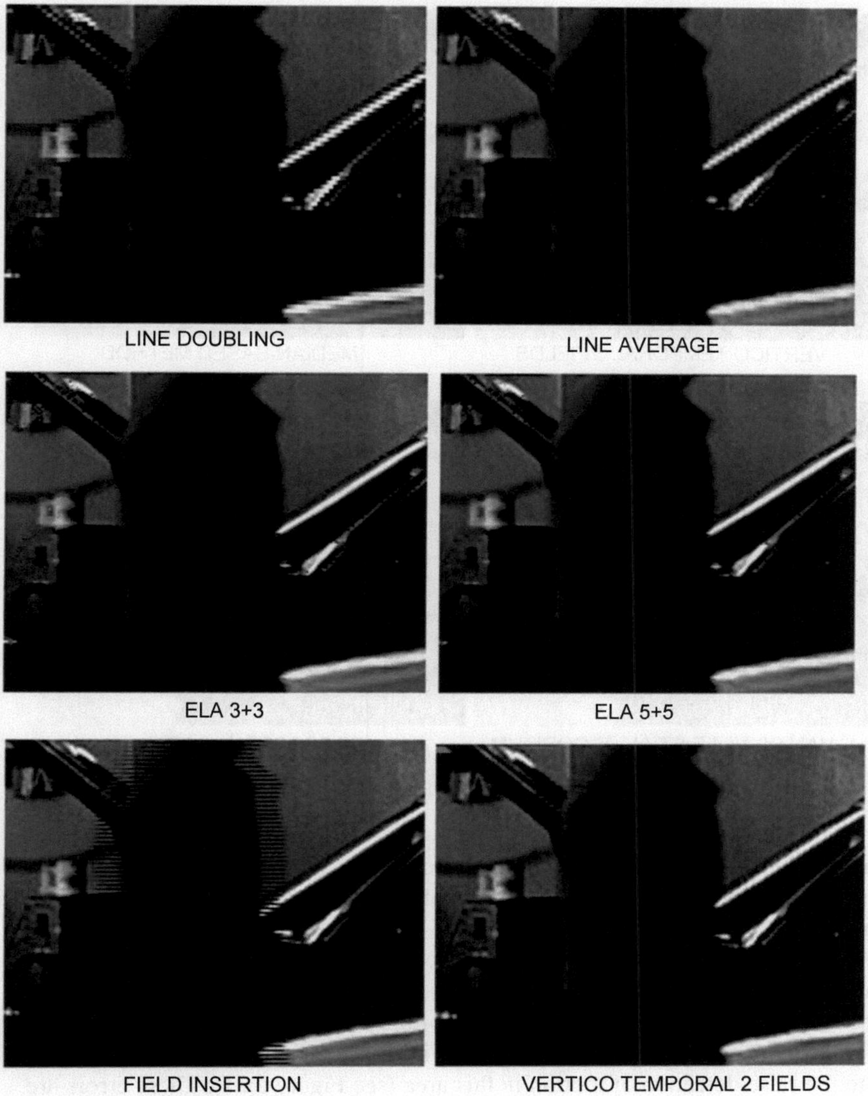

Fig. 2.34. De-interlaced images obtained with several de-interlacing techniques for the Fargo2 sequence.

Detailed areas of the de-interlaced images of the Fargo2 sequence are shown in Figures 2.34 and 2.35. In Figure 2.34, line doubling and line average can not reconstruct properly the clear edges of the image (see the boundaries of the black car). Both edge-adaptive methods de-interlace better the clear edges but they introduce some mistakes (black points in the gray hoods of the cars). 'Ghosting' effect introduced by field insertion is very emphasized in this sequence since the level of

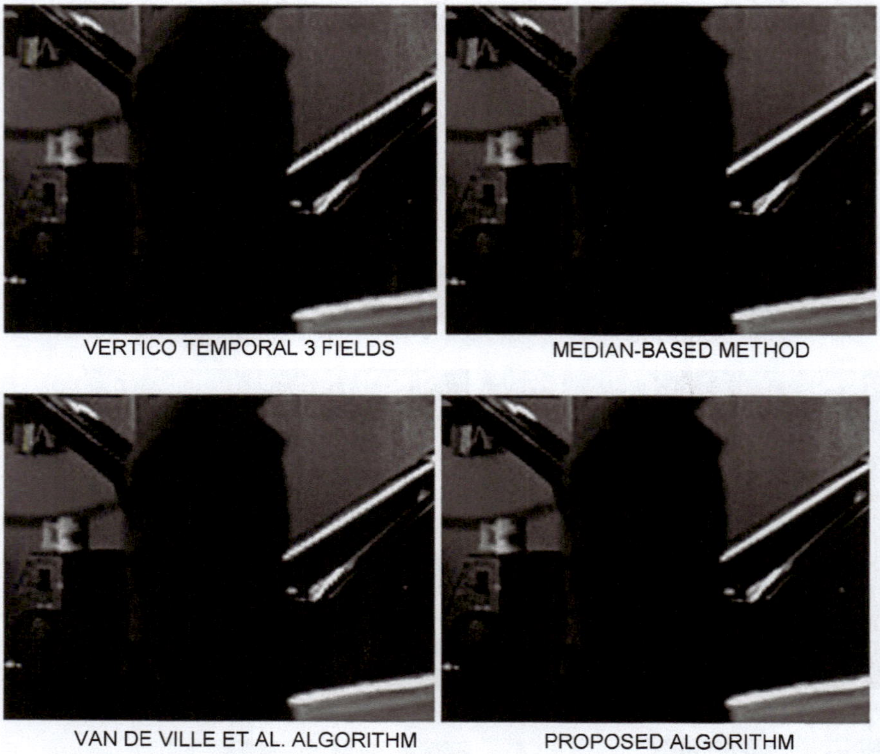

Fig. 2.35. De-interlaced images for the Fargo2 sequence.

motion in the man's body is very high. Vertico-temporal technique with two fields introduces significant mistakes in the interpolation of the man's body (white stains in the overalls' mechanic).

Vertico-temporal with three fields eliminates this problem but it can not interpolate the clear edges of the image as shown in Figure 2.35. The median-based method introduces small errors in the moving areas of the man's body and the Van de Ville et al. algorithm also presents a few mistakes in the reconstruction of edges. This is due to the false detection of motion in this area (see Figure 2.21). These errors are reduced by our approach since it detects a low level of motion in the background and, thus, it weights the spatial interpolation method (I_S) over the temporal method (I_T).

2.5 Conclusions

Between simple linear de-interlacing algorithms and complex motion-compensated ones, motion-adaptive algorithms represent a suitable midpoint. Efficiency of motion-adaptive algorithms is very much dependent on the quality of motion detection. Motion is usually a feature of part of the image and, hence, detection of motion is usually improved by processing images on a pixel by pixel basis, taking into account the neighborhood of each pixel. The other relevant issue when designing a motion-adaptive algorithm is to adapt the interpolator to the motion level measured. The first motion-adaptive algorithms proposed in the literature switched between simple temporal and spatial interpolators whenever the motion level increases over a threshold value, with the problem of selecting a suitable threshold. Van de Ville et al. employed fuzzy logic-based rules to cope with both issues: robust motion measurement and fuzzy selection of interpolators.

The fuzzy motion-adaptive algorithm proposed in this chapter reduces computational complexity of Van de Ville et al. proposal up to a percentage of 56% in terms of primitive operations. This improvement is a consequence of, firstly, using the bi-dimensional convolution proposed by Gutierrez et al. instead of 15 fuzzy rules with 5 antecedents to measure the motion level at each pixel and, secondly, using two simple complementary rules to implement a fuzzy selection between temporal and spatial interpolators instead of two non-complementary sets of 15 rules. In addition, the results obtained by our proposal (in terms of PSNR and visual inspection) after de-interlacing a wide battery of sequences of different formats are better than those obtained by Van de Ville et al. and other de-interlacing algorithms of similar complexity (such as vertico-temporal and median-based approaches). The Image Processing Toolbox of Matlab has been employed to perform all these extensive tests.

Open issues which are analyzed in the following chapter are how to select the size and weight values of the bi-dimensional convolution employed to measure the motion level as well as the number and type of fuzzy rules employed to adapt interpolators to motion.

Acknowledgements. The authors would like to express their gratitude to Dr. Julio Gutiérrez-Ríos, who introduced the bi-dimensional convolution techniques to measure the magnitude of motion.

References

1. Bock, A.M.: Motion-adaptive standards conversion between formats of similar field rates. Signal Processing: Image Communication 6(3), 275–280 (1994)
2. Jiang, H., Huu, D., Tinyork, E.: Motion adaptive deinterlacing. United States Patent (US 6,459,455) (October 2002)
3. Lin, W.-K., Lu, C.-Y.: Method for motion pixel detection. United States Patent (US 7,034,888) (April 2006)
4. Lin, W.-K., Lu, C.-Y.: Method for motion pixel detection with adaptive thresholds. United States Patent (US 7,242,435) (July 2007)
5. de Haan, G.: De-interlacing. In: de Haan, G. (ed.) Digital Video. Post Processing, September 2006, pp. 185–201. University Press Eindhoven (2006)
6. Van de Ville, D., Van de Wall, R., Philips, W., Lemahieu, I.: Motion adaptive de-interlacing using fuzzy logic. In: Proc. International Conference on Information Processing and Management of Uncertainty in Knowledge-Based Systems (IPMU), July 2002, pp. 1989–1996 (2002)

7. Van De Ville, D., Philips, W., Lemahieu, I.: Fuzzy-based motion detection and its application to de-interlacing. In: Fuzzy techniques in image processing. Studies in Fuzziness and Soft Computing. Physica-Verlag Editorial, Heidelberg (2000)
8. Gutiérrez-Ríos, J., Fernández-Hernández, F., Crespo, J.C., Trevino, G.: Motion adaptive fuzzy video de-interlacing method based on convolution techniques. In: Proc. of Information Processing and Management of Uncertainty in Knowledge-Based Systems (IPMU), Perugia (Italy) (July 2004)
9. Doyle, T., Looymans, M.: Progressive scan conversion using edge information. In: Chiariglione, L. (ed.) Signal Processing of HDTV, II, pp. 711–721. Elsevier Science Publishers, Amsterdam (1990)
10. Preliminary data sheet of Genesis gmVLDS, 8-bit Digital Video Line Doubler. version 1.0 (June 1996)
11. Weston, M.: Interpolating lines of video signals. Assignee: British Broadcasting Corporation, London, GB2. United States Patent Office US 4,789,893 December 6 (1988)
12. Haavisto, P., Neuvo, Y.: Motion adaptive scan rate up-conversion. Multidimensional Systems Signal Processing (3), 113–130 (1992)

Chapter 3
Design Options of the Fuzzy Motion-Adaptive Algorithm

Abstract. The aim of this Chapter is to analyze the design options of the fuzzy logic-based system for video de-interlacing introduced in Chapter 2, which combines a spatial and a temporal interpolator accordingly to the presence of motion. Particularly, the research work presented herein is focused on studying two topics: 1) the convolution mask used to calculate the input to the fuzzy system; and 2) the fuzzy rules employed by the system. Regarding convolution mask, the influence of considering several neighborhoods has been studied by de-interlacing the battery of video and film sequences. The results are evaluated in terms of the outcoming efficiency and computational simplicity, which means facilitating the implementation. Once the influence of the convolution mask is analyzed and one of the options is chosen, the structure of the fuzzy system used to carry out the de-interlacing task is studied. Specifically, the rule base is extended from the initial approach with two rules up to a set of more sophisticated systems, among which the most complex one works with five rules. The results obtained by de-interlacing the battery of sequences leads to the selection of the most convenient approach in terms of efficiency and complexity. As a consequence of the research work developed in this Chapter, an improved fuzzy-logic based approach for motion-adaptive de-interlacing is obtained. The new solution uses the most attractive proposal for the convolution mask and the most effective rule base among the alternatives presented herein. This Chapter is organized in three Sections. In Section 3.1, the options for the convolution mask are presented and their performances are studied. After selecting one of the convolution masks, the rule base and the parameters of the fuzzy systems are studied in Section 3.2. Finally, the conclusions of this Chapter are expounded in Section 3.3.

3.1 Convolution Mask Options

Since there are no 'a priori' constraints on the size of the convolution mask, our aim is to find out a solution that offers a good trade-off between complexity and de-interlacing quality.

Our study considers a set of convolution masks of different sizes so as to evaluate different spatio-temporal neighborhood of the current pixel in the calculus of the

P. Brox et al.: Fuzzy Logic-Based Algorithms for Video De-Interlacing, STUDFUZZ 246, pp. 85–103.
springerlink.com

CONVOLUTION MASKS FRAME DIFFERENCE MATRICES

$$C = \begin{bmatrix} 1 & 2 & 3 & 2 & 1 \\ 1 & 3 & 5 & 3 & 1 \\ 1 & 2 & 3 & 2 & 1 \end{bmatrix} \Rightarrow H = \begin{bmatrix} H_{(-2,-1,-1)} & H_{(-1,-1,-1)} & H_{(0,-1,-1)} & H_{(1,-1,-1)} & H_{(2,-1,-1)} \\ H_{(-2,0,0)} & H_{(-1,0,0)} & H_{(0,0,0)} & H_{(1,0,0)} & H_{(2,0,0)} \\ H_{(-2,1,-1)} & H_{(-1,1,-1)} & H_{(0,1,-1)} & H_{(1,1,-1)} & H_{(2,1,-1)} \end{bmatrix}$$

$$C1 = \begin{bmatrix} 1 & 1 & 2 & 1 & 1 \\ 1 & 2 & 3 & 2 & 1 \\ 2 & 3 & 4 & 3 & 2 \\ 1 & 2 & 3 & 2 & 1 \\ 1 & 1 & 2 & 1 & 1 \end{bmatrix} \Rightarrow H = \begin{bmatrix} H_{(-2,-2,0)} & H_{(-1,-2,0)} & H_{(0,-2,0)} & H_{(1,-2,0)} & H_{(2,-2,-1)} \\ H_{(-2,-1,-1)} & H_{(-1,-1,-1)} & H_{(0,-1,-1)} & H_{(1,-1,-1)} & H_{(2,-1,-1)} \\ H_{(-2,0,0)} & H_{(-1,0,0)} & H_{(0,0,0)} & H_{(1,0,0)} & H_{(2,0,0)} \\ H_{(-2,1,-1)} & H_{(-1,1,-1)} & H_{(0,1,-1)} & H_{(1,1,-1)} & H_{(2,1,-1)} \\ H_{(-2,2,0)} & H_{(-1,2,0)} & H_{(0,2,0)} & H_{(1,2,0)} & H_{(2,2,0)} \end{bmatrix}$$

$$C2 = \begin{bmatrix} 1 & 2 & 1 \\ 2 & 3 & 2 \\ 3 & 4 & 3 \\ 2 & 3 & 2 \\ 1 & 2 & 1 \end{bmatrix} \Rightarrow H = \begin{bmatrix} H_{(-1,-2,0)} & H_{(0,-2,0)} & H_{(1,-2,0)} \\ H_{(-1,-1,-1)} & H_{(0,-1,-1)} & H_{(1,-1,-1)} \\ H_{(-1,0,0)} & H_{(0,0,0)} & H_{(1,0,0)} \\ H_{(-1,1,-1)} & H_{(0,1,-1)} & H_{(1,1,-1)} \\ H_{(-1,2,0)} & H_{(0,2,0)} & H_{(1,2,0)} \end{bmatrix}$$

$$C3 = \begin{bmatrix} 1 & 2 & 1 \\ 2 & 4 & 2 \\ 1 & 2 & 1 \end{bmatrix} \Rightarrow H = \begin{bmatrix} H_{(-1,-1,-1)} & H_{(0,-1,-1)} & H_{(1,-1,-1)} \\ H_{(-1,0,0)} & H_{(0,0,0)} & H_{(1,0,0)} \\ H_{(-1,1,-1)} & H_{(0,1,-1)} & H_{(1,1,-1)} \end{bmatrix}$$

$$C4 = \begin{bmatrix} 0 & 1 & 0 \\ 1 & 2 & 1 \\ 0 & 1 & 0 \end{bmatrix} \Rightarrow H = \begin{bmatrix} H_{(-1,-1,-1)} & H_{(0,-1,-1)} & H_{(1,-1,-1)} \\ H_{(-1,0,0)} & H_{(0,0,0)} & H_{(1,0,0)} \\ H_{(-1,1,-1)} & H_{(0,1,-1)} & H_{(1,1,-1)} \end{bmatrix}$$

$$C5 = \begin{bmatrix} 1 \\ 2 \\ 1 \end{bmatrix} \Rightarrow H = \begin{bmatrix} H_{(0,-1,-1)} \\ H_{(0,0,0)} \\ H_{(0,1,-1)} \end{bmatrix}$$

Fig. 3.1. Convolution mask options.

motion level (see Figure 3.1). All of them assign the highest weight to the position placed on the current pixel while the values of the rest of weights are lower as the corresponding pixel is further from the current one. The idea is to reduce the influence of a neighbor in the measurement of the level of motion proportionally to its distance to the current pixel. This criterion, which seems logical, was confirmed quantitatively by comparing the convolution mask C1 with a 5x5 mask with all its elements 1. The results obtained with C1 were considerably better.

Since the convolution mask 'C' is normalized by the sum of its coefficients, the selection of their values is made to add a power of two as it can be corroborated in the matrices 'C' proposed in Figure 3.1. This eases the process of normalization by shifting the corresponding number of bits.

Besides the convolution mask employed in Chapter 2 (matrix 'C'), five possibilities have been considered in this study as shown in Figure 3.1. They range from a complex 5x5 matrix called C1, up to the simplest convolution mask, C5, which

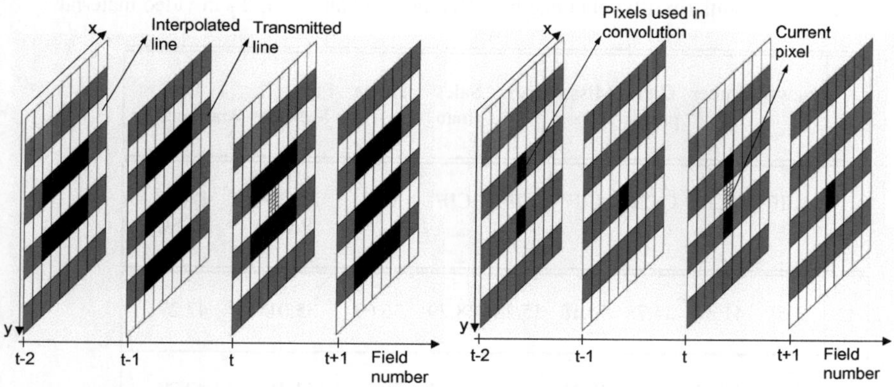

(a) Pixels used in matrix C1 (b) Pixels used in matrix C5

Fig. 3.2. The pixels used in C1 and C5 convolution masks are depicted in black.

only considers the influence of neighbors in the vertical direction. This direction has priority versus the horizontal one since de-interlacing is an interpolation process to reconstruct the vertical sampling density. A further simpler option is to measure the frame difference at only the current spatial position. However, this option was finally discarded to ensure that motion should involve several neighboring pixels in a picture, that is, pixels do not move separately. Therefore, the use of a neighborhood is very advantageous to avoid false cases of motion detection. This was corroborated by comparing the de-interlaced results obtained with this option and with the convolution mask C5. The results obtained with C5 were considerably better.

Each convolution mask is associated with a frame difference matrix as shown in Figure 3.1. To calculate the value of the bi-dimensional convolution, the following expression is applied:

$$motion(x,y,t) = \frac{\Sigma_a(\Sigma_b H_{a,b} C_{a,b})}{\Sigma_{a,b} C_{a,b}} \tag{3.1}$$

where $H_{a,b}$ are the elements of the frame difference matrix and $C_{a,b}$ are the elements of the convolution mask (one of the convolution masks C1, C2, C3, C4, and C5 shown in Figure 3.1).

The number of pixels involved in the calculation of the bi-dimensional convolution depends on the choice of the convolution mask. For instance, Figure 3.2(a) shows the thirty pixels (in black) used with the most complex convolution mask. In case that the simplest mask is chosen, only six pixels are evaluated, as shown in Figure 3.2(b).

3.1.1 Simulation Results

The performances of the alternatives proposed above are analyzed by de-interlacing the battery of video and film sequences. All the approaches use the same fuzzy logic-based system, which was presented in Chapter 2, that is, *line average* and *field*

Table 3.1. PSNR results (in dBs) obtained by different convolution masks in video material

	News	Mother	Car-phone	Missa	Paris	Sales-man	Trevor	Phantom Menace	Die AnotherDay
	QCIF	QCIF	QCIF	CIF	CIF	CIF	CIF	DVD PAL	DVD NTSC
C	37.51	41.87	34.78	40.18	35.28	38.29	36.69	35.01	42.27
C1	37.63	41.91	34.87	40.18	35.38	38.31	36.76	34.46	42.29
C2	37.61	41.81	34.86	40.17	35.36	38.31	36.73	34.71	42.23
C3	37.52	41.94	34.96	40.21	35.31	38.31	36.71	34.45	42.25
C4	37.48	41.89	35.03	40.24	35.38	38.31	36.68	34.33	42.22
C5	37.39	41.36	35.11	40.27	35.39	38.32	36.76	34.32	42.21

insertion have been used as spatial and temporal interpolators, respectively. Their difference is the bi-dimensional convolution employed to calculate 'motion'.

The PSNR results for the video sequences are presented in Table 3.1. Five sequences (Missa, Paris, Salesman, Trevor, and Die Another Day) seem to be rather independent of the convolution matrix employed (since the difference between the best and the worst cases are less than 0.12 dBs). The differences are also low in News and Carphone sequences (less than 0.25 and 0.34 dBs, respectively). Two video sequences, Mother and Phantom Menace, out of the nine sequences evaluated seem to be somewhat affected by the convolution mask (differences up to 0.58 and 0.69 dBs, respectively). The scarce influence of the convolution mask when de-interlacing video material is appreciated better by calculating the average value obtained with each convolution mask, as shown in Figure 3.3. There is no a convolution mask significantly better or worse than the others since the PSNR average difference between using the best (C2) and the worst (C5) is less than 0.07 dBs. The visual inspection of de-interlaced material neither reveals significant improvements of one convolution mask over the others.

The results achieved by de-interlacing film sequences are shown in Table 3.2. For the Tokyo sequence the PSNR values are quite similar. The differences in the Fire Rose and Fargo2 sequences are less than 0.34 and 0.38 dBs, respectively.

Table 3.2. PSNR results (in dBs) obtained by different convolution masks in film material

	Fire Rose	Fargo1	Fargo2	Tokyo	Bodyguard
	PAL TV	PAL TV	PAL TV	PAL TV	PAL TV
C	41.14	37.33	42.54	37.71	40.28
C1	41.08	37.05	42.91	37.72	40.14
C2	41.05	37.05	42.61	37.57	40.02
C3	41.18	37.35	42.66	37.66	40.51
C4	41.24	37.52	42.67	37.72	40.66
C5	41.38	37.41	42.63	37.71	40.58

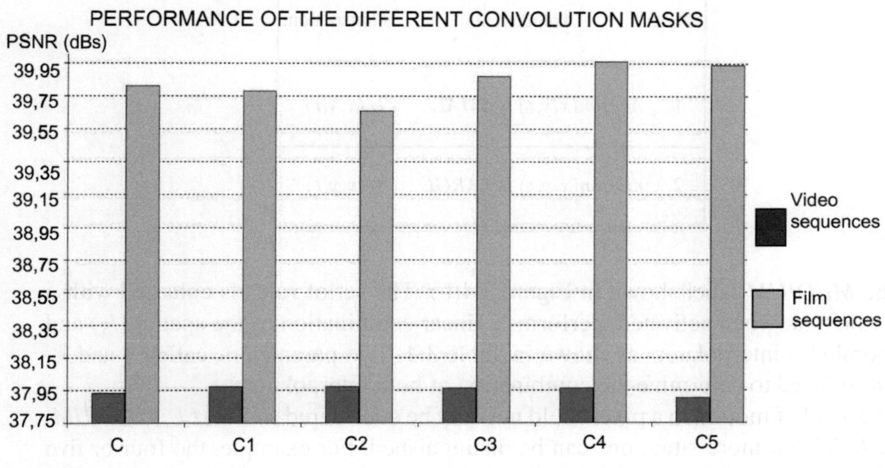

Fig. 3.3. The average value of PSNR results obtained with the different convolution masks.

Convolution masks seem affect only to the Fargo1 (up to 0.47 dBs) and the Bodyguard (up to 0.64 dBs) sequences.

Once again the results are more understandable by representing the PSNR average value obtained for each convolution mask, as shown in Figure 3.3. In film material, the difference between using the best (C4) or the worst (C2) convolution mask is 0.15 dBs. Anyway, the visual inspection of de-interlaced images neither discovers significant advantages of one convolution mask over the others.

Something interesting to remark is that the best average results in film material are obtained with the simple convolution masks (C4 and C5) while the worst results are obtained with convolution mask C2. On the contrary, C2 is the best mask and C5 is the worst one for video material.

All these simulation results lead us to think that the six convolution masks presented in this study provide a similar performance and, therefore, the simplest mask, C5 has been selected for the following study.

3.2 Rule Base Options

The approach presented in Chapter 2 uses a fuzzy inference system with two rules (see Table 3.3), where the concepts *SMALL* and *LARGE* motion are represented by the fuzzy sets shown in Figure 3.4(a). Nevertheless, the interpolation capacity of fuzzy logic could be further exploited by considering the possibility of extending the number of labels employed to discriminate the level of motion and, consequently, the number of rules. In this sense, it is possible to define a new fuzzy set represented

Table 3.3. De-interlacing based on 2 fuzzy rules

Rule	Antecedents	Consequent
1.	$motion(x,y,t)$ is *SMALL*	$I_T(x,y,t)$
2.	$motion(x,y,t)$ is *LARGE*	$I_S(x,y,t)$

by the *MEDIUM* label shown in Figure 3.4(b). The set of rules is enlarged with a new rule that, when activated, performs a linear combination of the spatial (I_S) and temporal (I_T) interpolators as shown in Table 3.4. Two parameters, called γ and λ, are introduced to determine the combination of both interpolators.

The level of motion in a pixel could not only be considered as *SMALL*, *MEDIUM*, or *LARGE*, but more situations can be distinguished. For example, the four or five labels represented in Figure 3.5 and Figure 3.6 could be employed. The use of more fuzzy sets directly implies an extension of the rule base. Table 3.5 shows the four 'if-then' rules considered when four labels are used. Table 3.6 shows a rule base with five rules, which is the largest system considered in this study. The consequents of the new rules establish new linear combinations between I_S and I_T interpolators.

Since the increment of the rule base complexity means a higher number of resources required to implement them, our objective is to select a balanced solution

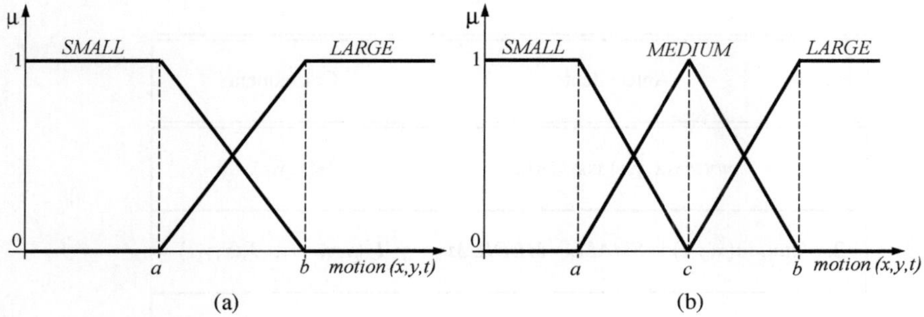

Fig. 3.4. Two and three fuzzy sets for the motion level.

Table 3.4. De-interlacing based on 3 rules

Rule	Antecedents	Consequent
1.	$motion(x,y,t)$ is *SMALL*	$I_T(x,y,t)$
2.	$motion(x,y,t)$ is *MEDIUM*	$\gamma\, I_T(x,y,t)+\lambda\, I_S(x,y,t)$
3.	$motion(x,y,t)$ is *LARGE*	$I_S(x,y,t)$

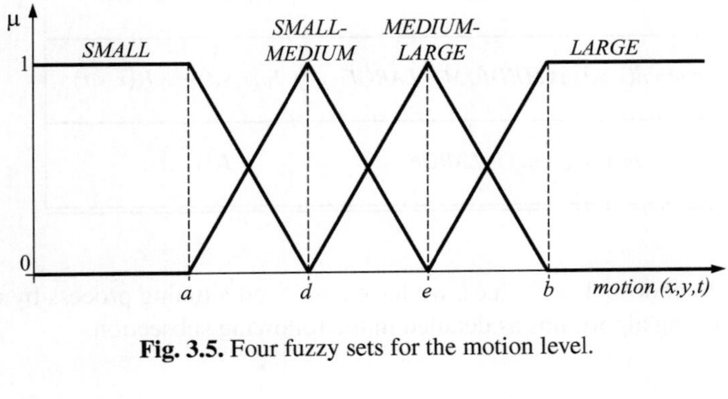

Fig. 3.5. Four fuzzy sets for the motion level.

between complexity and quality. Therefore, the introduction of more rules should be justified by a significant improvement of the de-interlacing performance.

Heuristic knowledge does not provide enough information neither to fix the constant values γ, γ_1, γ_2 and λ, λ_1, λ_2 of the rules' consequents, nor to determine the values a, b, c, d, and e, that describe the five possible linguistic labels (see Figure

Table 3.5. De-interlacing based on 4 rules

Rule	Antecedents	Consequent
1.	$motion(x,y,t)$ is $SMALL$	$I_T(x,y,t)$
2.	$motion(x,y,t)$ is $SMALL-MEDIUM$	$\gamma_1\, I_T(x,y,t)+\lambda_1\, I_S(x,y,t)$
3.	$motion(x,y,t)$ is $MEDIUM-LARGE$	$\gamma_2\, I_T(x,y,t)+\lambda_2\, I_S(x,y,t)$
4.	$motion(x,y,t)$ is $LARGE$	$I_S(x,y,t)$

Table 3.6. De-interlacing based on 5 rules

Rule	Antecedents	Consequent
1.	$motion(x,y,t)$ is $SMALL$	$I_T(x,y,t)$
2.	$motion(x,y,t)$ is $SMALL-MEDIUM$	$\gamma_1\, I_T(x,y,t)+\lambda_1\, I_S(x,y,t)$
3.	$motion(x,y,t)$ is $MEDIUM$	$\gamma\, I_T(x,y,t)+\lambda\, I_S(x,y,t)$
4.	$motion(x,y,t)$ is $MEDIUM-LARGE$	$\gamma_2\, I_T(x,y,t)+\lambda_2\, I_S(x,y,t)$
5.	$motion(x,y,t)$ is $LARGE$	$I_S(x,y,t)$

3.6). In order to choose these values, we have employed a tuning process by using supervised learning algorithms as detailed in the following subsection.

3.2.1 Tuning of Membership Function and Consequent Parameters

The approach we have followed to select good values for the membership function parameters and the constants in the rules' consequents has been to employ a set

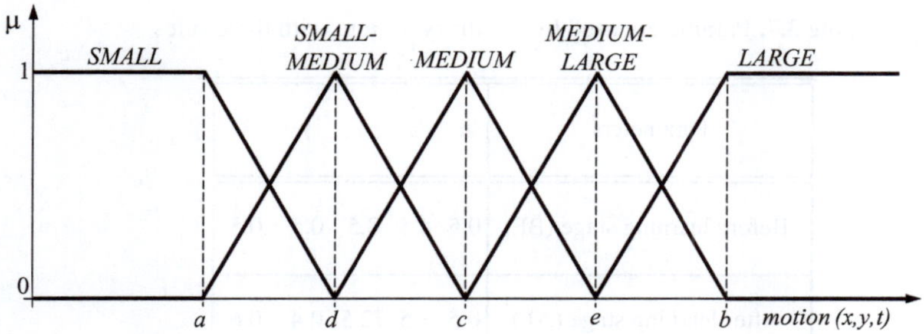

Fig. 3.6. Five fuzzy sets for the motion level.

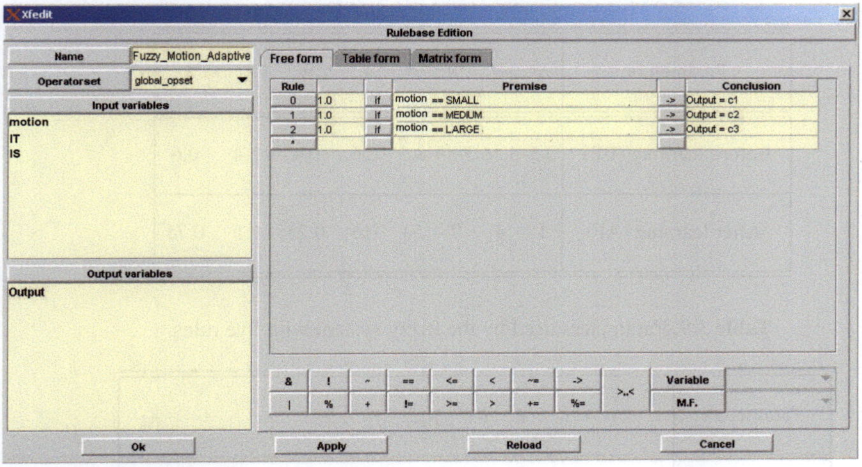

Fig. 3.7. Description of the three-rule fuzzy system in Xfuzzy 3.1.

of data from progressive images, so as to minimize an error function by applying supervised learning algorithms.

In order to use the CAD tool *xfsl* of Xfuzzy 3.1 environment, which allows applying a wide set of supervised learning algorithms, the above introduced fuzzy systems have to be described by XFL3, the specification language of Xfuzzy 3.1. Considering 'I_T', 'I_S', and '*motion*' as the input variables, the fuzzy systems have been described as first-order Takagi-Sugeno systems since the rules' consequents are linear functions of the inputs 'I_T' and 'I_S'. For instance, the description of the fuzzy system with three rules by using the graphical user interface of the CAD tool *xfedit* is illustrated in Figure 3.7.

Once the fuzzy system has been described in XFL3, the process to perform the tuning stage is ready to be launched. This is performed with the tool *xfsl* by using the well-known Marquardt-Levenberg algorithm as supervised learning algorithm. The set of parameters enabled to participate in the tuning process is composed by *a*,

Table 3.7. Parameters used by the fuzzy system with three rules

Parameters	a	c	b	γ	λ
Before learning stage (BF)	0.5	4.5	8.5	0.5	0.5
After learning stage (AF)	0.5	8.5	72.5	0.4	0.6

Table 3.8. Parameters used by the fuzzy system with four rules

Parameters	a	d	e	b	γ_1	γ_2	λ_1	λ_2
Before learning (BF)	0.5	3.16	5.83	8.5	0.6	0.4	0.4	0.6
After learning (AF)	1	4	7	54	0.5	0.25	0.5	0.75

Table 3.9. Parameters used by the fuzzy system with five rules

Parameters	a	d	c	e	b	γ_1	γ	γ_2	λ_1	λ	λ_2
Before learning (BF)	0.5	2.5	4.5	6.5	8.5	0.7	0.5	0.3	0.3	0.5	0.7
After learning (AF)	1	2.3	4.2	12	64	0.78	0.4	0.2	0.22	0.6	0.8

b, c, γ, and λ for the fuzzy system with three rules. For the fuzzy systems with four and five rules the set is increased to include the parameters: d, e, γ_1, γ_2, λ_1, and λ_2.

Before learning, the fuzzy systems with three, four, and five rules use the set of parameters shown in Tables 3.7, 3.8 and 3.9, respectively. As can be seen in these tables, the values of all the initial proposals are 0.5 for the parameter a and 8.5 for the parameter b, which correspond to the initial values proposed by Van de Ville et al. for the *SMALL* and *LARGE* membership functions [1]. The values to determine the intermediate fuzzy sets are initially located in equidistant intervals between the minimum and maximum values. The criteria used to establish the initial values for the weights of the spatial and temporal interpolators are the following:

- The sum of the weights for both interpolators has to be equal to 1.
- Equal values of the weights are selected if *motion* is $MEDIUM$.
- A higher value for the weight of the spatial interpolator is used if *motion* is $MEDIUM-LARGE$. On the contrary, if *motion* is $SMALL-MEDIUM$ the weight of the temporal interpolator is higher than the one chosen for the spatial interpolator.

The values obtained after applying supervised learning algorithms are shown in Tables 3.7, 3.8 and 3.9. The most relevant result is the increase of the parameter *b* in all the fuzzy systems up to a numerical value higher than 50. The values of weights for the spatial and the temporal interpolator in the rules' consequents are also modified. Although the sum of the weights for the spatial and temporal interpolators is not constrained to be 1, the learning process converges to values that always fulfill the condition $\gamma_i+\lambda_i=1$, with $\{i=1,\ 2\ or\ nothing\}$.

3.2.2 Simulation Results

Performance of the different fuzzy systems (with 3, 4, and 5 rules) has been evaluated and compared by de-interlacing the battery of video and film sequences. All the new fuzzy systems use the same input signal, that is, the bi-dimensional convolution with the convolution mask *C*5, and the same spatial (I_S) and temporal (I_T) interpolators, *line average* and *field insertion*, respectively. The results obtained with the new design options of the fuzzy system are also compared with the initial approach presented in Chapter 2.

In addition, to show up the efficiency of using fuzzy sets to represent the antecedents of the rules, the proposed fuzzy systems have been also compared with similar ones that employ crisp definitions. The procedure to select the crisp values is illustrated in Figure 3.8 for the fuzzy system with three rules. The same philosophy is applied to obtain the crisp sets of the four- and five-rule systems. All the resulting crisp systems use the parameters obtained after learning to define the antecedents (*a*, *b*, *c*, *d*, *e*) and the consequents (λ, λ_1, λ_2, γ, γ_1, γ_2).

Table 3.10 shows the results obtained by the different options when de-interlacing video sequences. In order to facilitate the interpretation of these results, they have been displayed graphically (see Figures 3.9, 3.10, and 3.11).

The aim of the graph in Figure 3.9 is to compare the performance of the fuzzy systems before and after the learning stage. The superior part of the bars in the diagram illustrates the improvements achieved by the systems that have been defined by supervised learning algorithms. The number inside each bar indicates the number of rules of each system.

As can be corroborated in Figure 3.9, the AL systems obtain the highest PSNR values except for the 3-rule system in the Carphone sequence. The increase of the PSNR value AL is of 1 dB more or less for five sequences (News, Missa, Paris,

Table 3.10. PSNR results (in dBs) for different rule bases (video material)

	News	Mother	Car-phone	Missa	Paris	Sales-man	Trevor	Phantom Menace	Die AnotherDay
	QCIF	QCIF	QCIF	CIF	CIF	CIF	CIF	DVD PAL	DVD NTSC
Proposal in Chapter 2	37.51	41.87	34.78	40.18	35.28	38.29	36.69	35.01	42.27
3 rules (BL)	37.59	41.23	36.64	40.21	34.67	38.14	36.22	33.29	41.86
3 rules (AL)	38.28	41.35	35.05	40.51	35.77	38.45	37.29	34.39	42.35
3 rules CRISP	37.53	40.03	33.52	39.89	35.38	37.74	36.24	32.11	42.15
4 rules (BL)	37.36	41.04	34.76	39.81	34.78	38.17	36.21	34.26	41.81
4 rules (AL)	38.61	41.22	34.95	40.19	35.41	38.37	36.78	34.91	41.99
4 rules CRISP	37.47	40.91	34.77	39.66	35.02	38.03	36.27	34.06	41.79
5 rules (BL)	37.33	41.08	34.89	38.82	34.72	38.17	36.51	34.32	41.71
5 rules (AL)	38.08	41.25	34.98	40.29	35.92	38.21	37.16	34.96	41.98
5 rules CRISP	37.56	41.02	34.78	39.89	35.01	38.01	36.47	34.11	41.75

Trevor, and Phantom Menace). The improvements are inferior to 0.5 dBs for three sequences (Mother, Salesman, and Die Another Day).

Figure 3.10 shows a graph to illustrate the advantages of using fuzzy instead of crisp definitions for the fuzzy systems with three, four, and five rules. As can be seen, all the sequences except one (Die Another Day) are clearly affected by

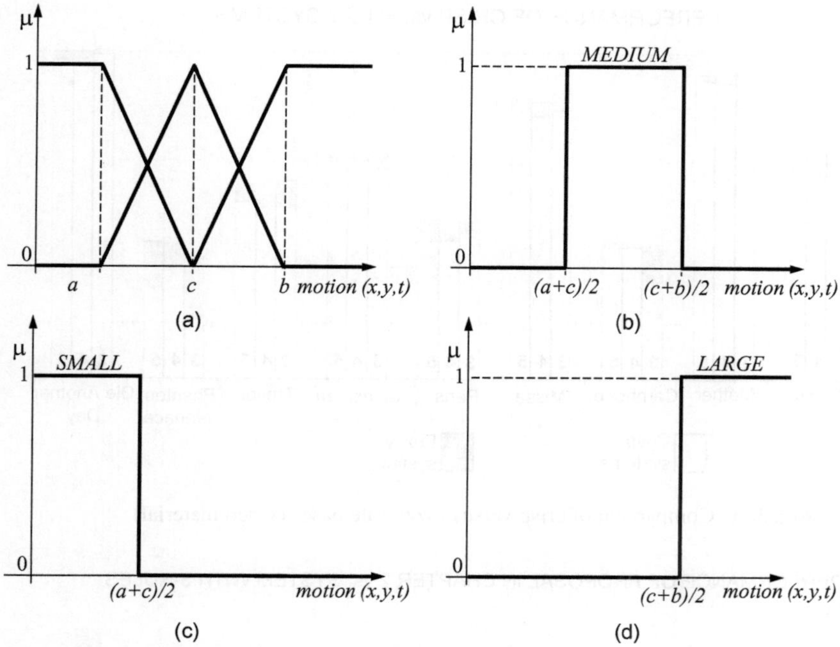

Fig. 3.8. Fuzzy membership functions versus crisp ones.

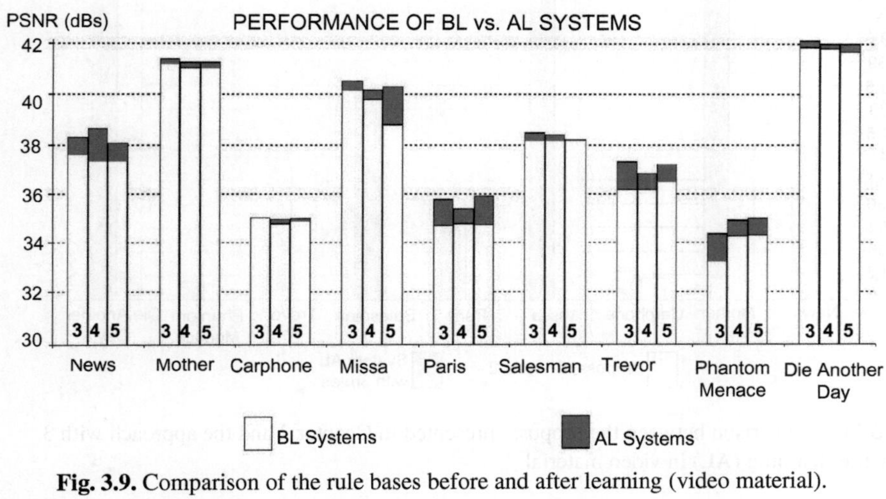

Fig. 3.9. Comparison of the rule bases before and after learning (video material).

using crisp instead of fuzzy definitions in case that the rule base contains three rules. The differences between the crisp and fuzzy approaches are reduced when the number of rules increases in six of the nine sequences of the battery (News, Mother, Missa, Salesman, Trevor, and Phantom Menace). As mentioned above, the

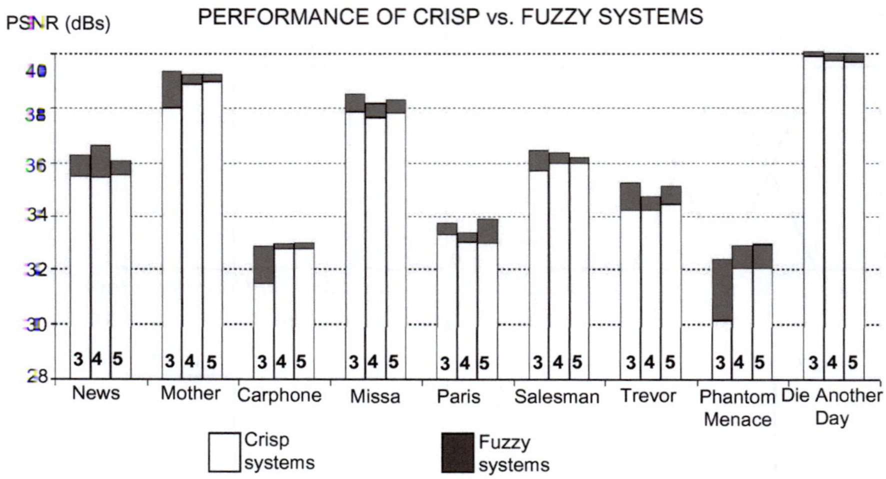

Fig. 3.10. Comparison of crisp versus fuzzy rule bases (video material).

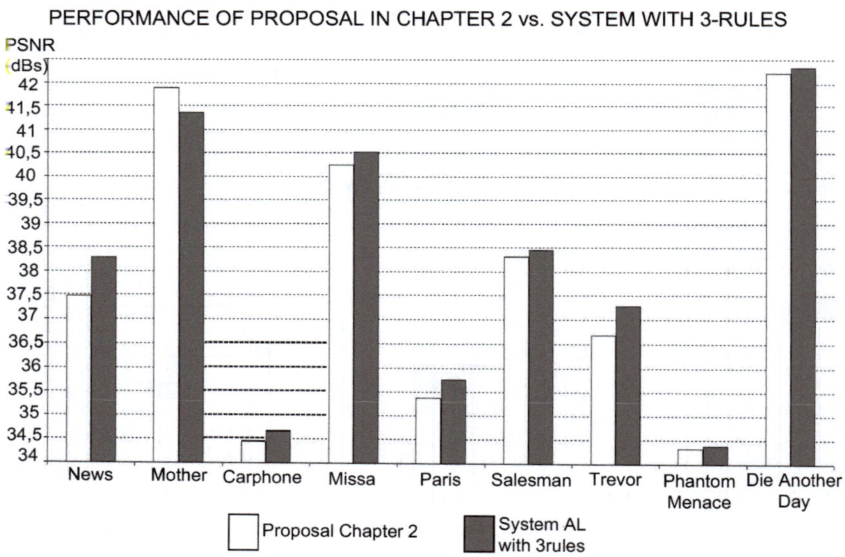

Fig. 3.11. Comparison between the proposal presented in Chapter 2 and the approach with 3 rules after learning (AL) in video material.

crisp labels are defined by using the tuned membership function and consequent parameters. Some of the simulations were initially carried out by using parameters BL but their performances were really poor in the majority of the sequences. This makes evident the critical influence of the parameter values in the design of the crisp systems.

Table 3.11. PSNR results (in dBs) for different rule bases (film material)

	Fire Rose	Fargo 1	Fargo 2	Tokyo	Bodyguard
	PAL TV	PAL TV	PAL TV	PAL TV	PAL TV
Proposal in Chapter 2	41.14	37.33	42.54	37.71	40.28
3 rules (BL)	39.79	36.48	42.29	36.92	37.69
3 rules (AL)	39.91	36.74	42.36	38.18	40.79
3 rules CRISP	39.47	36.56	40.66	37.82	35.07
4 rules (BL)	39.95	36.83	42.17	36.91	39.29
4 rules (AL)	40.05	37.12	42.22	37.49	39.96
4 rules CRISP	39.93	37.03	40.77	37.29	38.22
5 rules (BL)	39.95	36.83	42.02	36.92	38.75
5 rules (AL)	40.19	37.17	42.26	37.48	39.97
5 rules CRISP	40.02	36.92	41.82	37.26	37.02

The graph in Figure 3.11 compares the algorithm presented in Chapter 2 with a fuzzy system of 3 rules AL in video material. This comparison is clearly favorable for the latter one in all the sequences except for the Mother sequence. The poor performance of the Mother sequence is explained due to the features of this sequence. It contains areas of the frames with extreme cases of motion, that is, areas of the background that are totally static and zones where there are sudden and fast

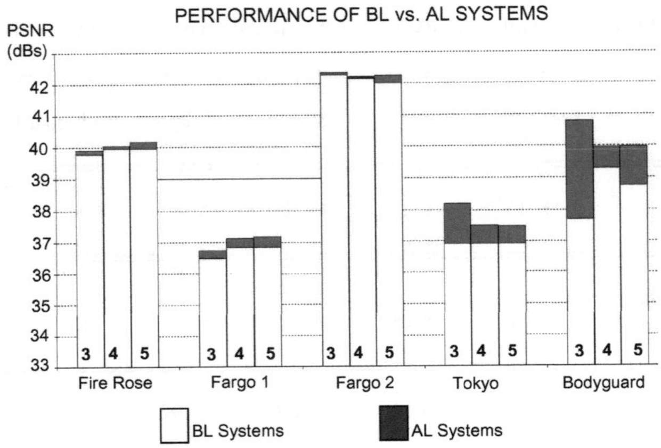

Fig. 3.12. Comparison of rule bases before and after learning (film material).

movements. Since these situations are better represented by two labels, the new fuzzy system with 3 rules gets worse results.

Among the different design options of the fuzzy system presented in Figures 3.9, 3.10, and 3.11, the fuzzy system with 3 rules AL offers the most attractive solution since it presents a good trade off between complexity and PSNR results.

The influence of the rule base has been also analyzed by de-interlacing film sequences. The results in Table 3.11 show the average PSNR obtained after de-interlacing the five film sequences. The results are again more apparent in bar diagrams. For instance, the advantages of learning the parameters are shown in Figure 3.12. The use of the tuned parameters allows to improve the PSNR results in all the film sequences. Particularly, the improvements are clearly significant in two of the five sequences (Tokyo and Bodyguard). All the training data employed by the supervised learning algorithm come from video material. Hence, it is interesting that the learned systems also perform better than non-learned systems with film material. This shows that learning does not overfit training data but allows generalization.

The techniques with crisp systems are compared to fuzzy systems in the bar diagram of Figure 3.13. Particularly, the improvements are evident in two sequences: Fargo2 and Bodyguard.

Figure 3.14 compares the system presented in Chapter 2 with the fuzzy system with 3 rules AL. The results are worse for the new fuzzy system in three of the five film sequences. This may lead to think that the new 3-rule fuzzy system is inefficient. However, this conclusion is wrong since the PSNR average values of film sequences are deceitful. To corroborate this, let us observe the graph in Figure 3.15 where the MSE result for Fargo2 sequence is shown field by field. Although fifty fields were analyzed, the graph in Figure 3.15 only includes the results from the 6th to the 44th field so as to distinguish properly the differences among both proposals: the extension of the rule base from two to three rules introduces improvements in odd fields of the sequence but the results get worse in even fields. These results make

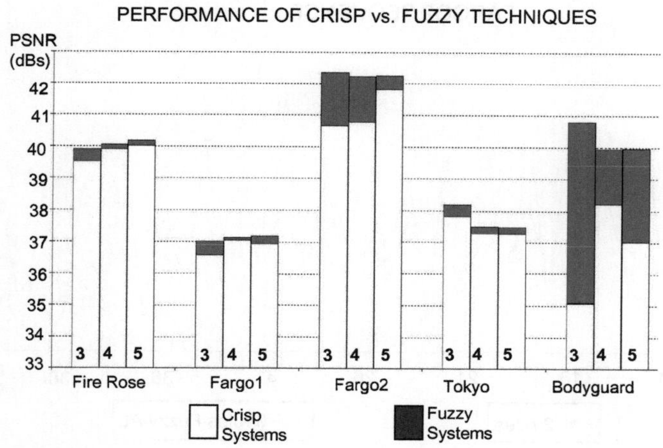

Fig. 3.13. Comparison of crisp versus fuzzy rule bases (film material).

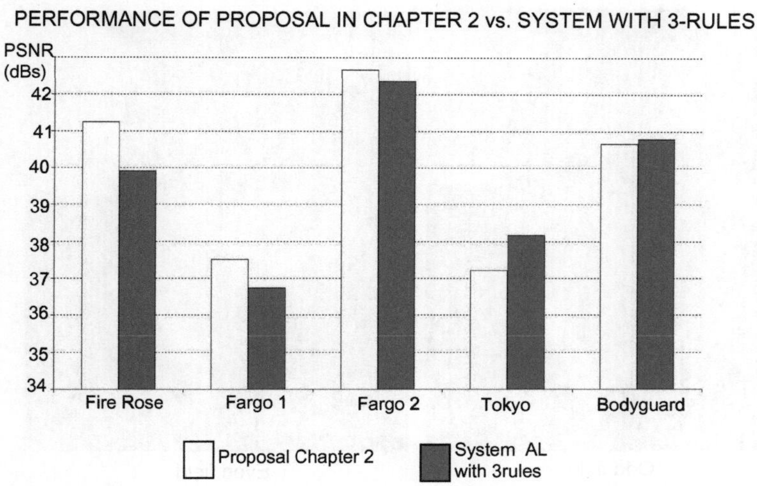

Fig. 3.14. Comparison between the proposal presented in Chapter 2 and the approach with 3 rules after learning (AL) in film material.

evident that the approach with 3 rules AL gives more importance to the temporal interpolator (I_T), and hence this improves the results in odd fields where the previous field agrees with the current one. However, the results are considerably worse in even fields where the previous field is different from the current one. This behavior of Fargo2 sequence is also shared by the rest of film sequences since all of them are derived from a 2:2 pull-down process. For instance, the loss of quality in an even de-interlaced field can be observed in Figure 3.16(right) for the Bodyguard

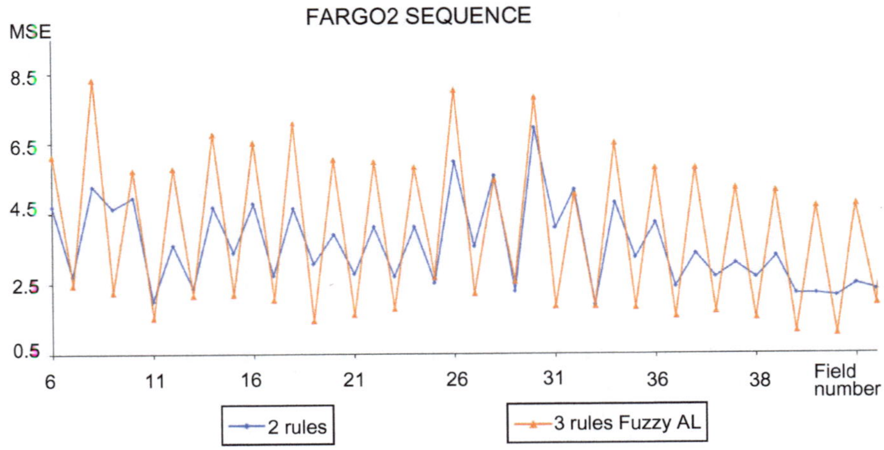

Fig. 3.15. MSE values for Fargo2 sequence field by field.

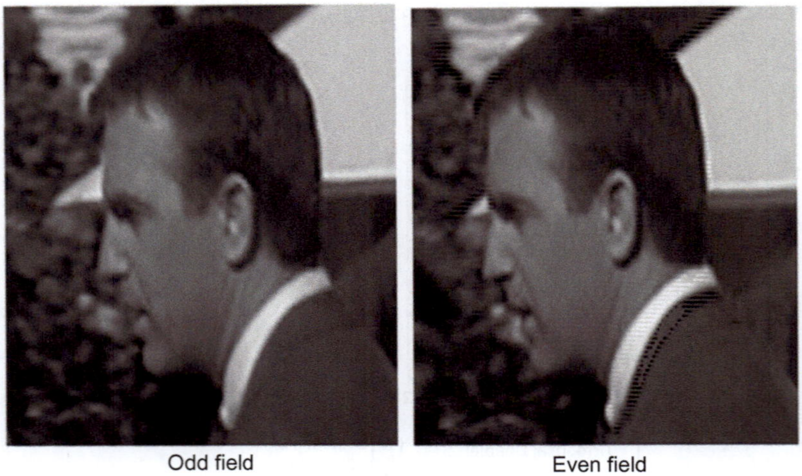

Fig. 3.16. Even and odd de-interlaced images for the Bodyguard sequence.

sequence. Since the previous field does not agree with the current one, the system with 3 rules, which gives more importance to temporal interpolator (field insertion) produces more ghosthering effects. On the contrary, Figure 3.16(left) shows the quality improvement in an odd de-interlaced field. Since the previous field agrees with the current one, the system with the 3 rules performs better.

After analyzing the results of de-interlacing film and video sequences, the algorithm that uses the fuzzy system with 3 rules AL gives the best results among all the alternatives. Furthermore, it also improves the performance of the algorithm presented in Chapter 2 in the majority of sequences. As conclusion, the proposal with the convolution mask C5 and a rule base with 3 rules AL is chosen to carry out motion-adaptive de-interlacing.

The resources required to implement the new algorithm are basically the same in terms of field memories and line buffers. The implementation of the new algorithm reduces the number of registers since the bi-dimensional convolution only works with pixels in vertical direction. Although the inclusion of the third rule in the rule base slightly complicates the system, the simplification of the convolution mask makes that the overall number of primitive operations is reduced from 45 to 26.

3.3 Conclusions

After analyzing extensively the performance of 6 different convolution masks in the bi-dimensional convolution that measures the motion level, the conclusion is that they do not have a significant influence on the de-interlaced results provided that two vertical neighbors are at least considered and the weight of the current pixel has always the highest value. Hence, the simplest mask is selected for the subsequent analysis.

Selection of the adequate fuzzy rules to employ involves choosing the number of rules as well as the parameters of the rules. Since given a fixed number of rules, the values of the rule parameters to analyze could be enormous, a supervised learning algorithm that minimizes the mean square error between a set of data corresponding to progressive and de-interlaced results of different sequences has been employed. The use of the CAD tool xfsl from the environment Xfuzzy 3 has facilitated the application of this methodology. Rule bases whose parameters are tuned by this way always improve their performance. Another important conclusion is that fuzzy rules always perform better than crisp rules. Regarding the number of rules, extensive results show that using more rules does not improve significantly the de-interlacing and even, sometimes, degrades the performance. Hence, three rules have been selected as the best option.

Once convolution masks and rule bases have been explored, the next way to improve de-interlacing will be to improve the spatial and temporal interpolators employed. Line average as spatial interpolator introduces an unwanted staircase effect on the de-interlaced edges (especially annoying in diagonal edges that move slowly in the vertical direction). Hence, a better spatial interpolator is introduced in Chapter 4. Field insertion as temporal interpolator has been proven (especially in film sequences) to be a poor solution. Hence, a better temporal interpolation is introduced in Chapter 5.

Reference

1. Van De Ville, D., Philips, W., Lemahieu, I.: Fuzzy-based motion detection and its application to de-interlacing. In: Fuzzy techniques in image processing. Studies in Fuzziness and Soft Computing. Physica-Verlag Editorial, Heidelberg (2000)

Chapter 4
Fuzzy Motion-Adaptive De-Interlacing with Edge-Adaptive Spatial Interpolation

Abstract. The quality of the motion-adaptive de-interlacing algorithm developed in previous Chapters can be increased by improving the performance of the spatial interpolator. Until now, *line average* has been used but its results are poor in the reconstruction of areas with edges. Since the human visual system is specially sensitive to the edges of an image, a correct interpolation of them has a vital importance to achieve a positive visual inspection. Hence, although an interpolated image obtained by *line average* could feature an overall PSNR value superior to the PSNR result obtained by an edge-dependent algorithm, human appreciation would usually prefer de-interlaced pictures obtained with edge-dependent algorithms. Let us illustrate the advantages of using an edge-adaptive algorithm by examining an image with a high number of clear defined edges as shown in Figures 4.1 and 4.2. The reconstruction of clear, sharp, and high-contrast edges are significantly improved by using a conventional edge-dependent algorithm such as ELA (Edge-based Line Average) because *line average* introduces a blurred effect. Edge-adaptive de-interlacing algorithms explore a neighborhood of the current pixel to extract information about the edge orientation [1], [2]-[9]. Once determined the direction with the major probability of an edge, the luminance is interpolated in that direction. The first proposal following this approach, which was called ELA [1], works with 3+3 pixels in the upper and lower lines. It performs well when the edge direction agrees with the maximum correlation, but otherwise introduces mistakes and degrades the image quality. These mistakes usually appear when the edges are not clear, the image is corrupted by noise, there is a high number of details, etc. In addition, ELA lacks the ability to detect horizontal edges. Several proposals have been presented recently in the literature to avoid the above shortcomings of the ELA algorithm [2]-[9]. The approaches reported in [2] and [3] focus on enhancing the reconstruction of horizontal edges. In [2], line doubling and an edge-adaptive de-interlacing algorithm based on ELA are combined to interpolate horizontal edges. In [3], an edge-based line average is proposed, which uses the 3+3 pixels of ELA and the interpolated pixel value that was calculated previously. Other algorithms use a larger neighborhood to get more information about the possible edge direction. For instance, the algorithm presented in [4] consists of a modified ELA module and a contrast enhancement module. The ELA module increases the processing window up to 5+5

P. Brox et al.: Fuzzy Logic-Based Algorithms for Video De-Interlacing, STUDFUZZ 246, pp. 105–142.
springerlink.com

pixels whereas the second module reduces detail losses due to the interpolation. The neighborhood is enlarged up to 6+6 taps in [5], 7+7 taps in [6], [7], 11+11 taps in [8], and 34+34 taps in [9]. A higher number of pixels in the aperture provides much information about edges but at expense of increasing the algorithm complexity. The algorithms presented in this Chapter are inspired by the ELA scheme and include a simple fuzzy system that models heuristic knowledge to ensure the presence of edges. Thus, they have been named Fuzzy-ELA algorithms. This Chapter is organized as follows: Section 4.1 describes the 'Basic Fuzzy-ELA' algorithm and its performance when de-interlacing the battery of sequences presented in the previous Chapters. Further modifications of the 'Basic Fuzzy-ELA' algorithm are presented

Fig. 4.1. (a) A progressive image that contains sharp and clearly defined edges. Zoom of images obtained after de-interlacing by (b) *line average*, and (c) ELA algorithm.

(a) (b) (c)

Fig. 4.2. (a) An area with clear edges in the progressive image of Figure 4.1. De-interlaced results obtained by (b) *line average*, and (c) ELA algorithm.

and analyzed in Section 4.2. After comparing these approaches, one of them is selected as spatial interpolator to be used in our fuzzy motion-adaptive algorithm in Section 4.3. Finally, some conclusions are expounded in Section 4.4.

4.1 Basic Fuzzy-ELA Algorithm

In order to describe our basic proposal that is inspired by the conventional ELA, let us simplify the notation used in previous Chapters. The new interpolated pixel value, $I_P(x,y,t)$, is expressed herein as X. ELA algorithm interpolates X by analyzing the luminance differences of the 3+3 pixels in the upper and lower lines (see Figure 4.3). ELA looks for the most possible edge direction and then applies *line average* along the selected direction. The pseudo-code of the ELA algorithm is as follows:

$$\begin{aligned} &a = |A-F| \quad b = |B-E| \quad c = |C-D| \\ &if \;\; min(a,b,c) = a \rightarrow X = (A+F)/2 \\ &elseif \;\; min(a,b,c) = c \rightarrow X = (C+D)/2 \\ &else \;\; \rightarrow X = (B+E)/2 \end{aligned} \tag{4.1}$$

ELA achieves suitable results if the maximum correlation agrees with the presence of an edge, which usually happens to clear edges. Nevertheless, problems appear since the maximum correlation does not always indicate the direction of an edge. Figure 4.4 illustrates several cases where ELA introduces errors. The letter shown in Figure 4.4(b), which is a detailed area of the progressive image in Figure 4.4(a), shows different examples of edges: clear edges, unclear edges, and ambiguous situations for ELA algorithm. If ELA is applied to interpolate the central pixel of the middle row in the examples of Figure 4.4(b), the results obtained are shown in Figure 4.5. As corroborated in this image, ELA fails in the interpolation of unclear and ambiguous edges.

The mistakes introduced by ELA can also be observed after de-interlacing the image in Figure 4.4(a). ELA performs very well in clear edges, such as the black and white stripped lines shown in Figure 4.6(a). However, its performance is poor in the calendar area where there are many unclear and ambiguous situations (see Figure 4.6(b)).

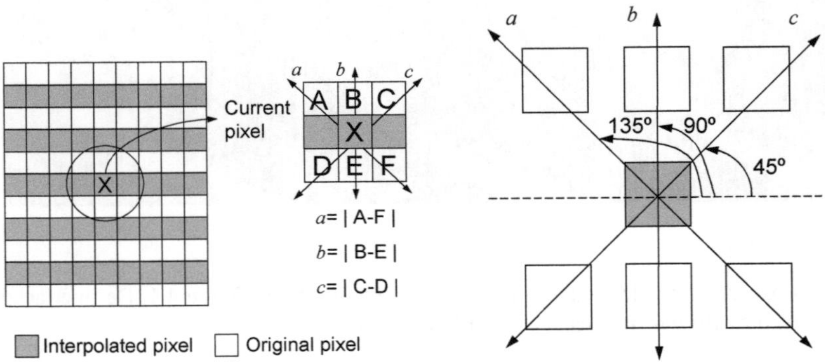

(a) ELA scheme (b) 3+3 Aperture ELA

Fig. 4.3. ELA algorithm with 3+3 pixels from the upper and lower lines.

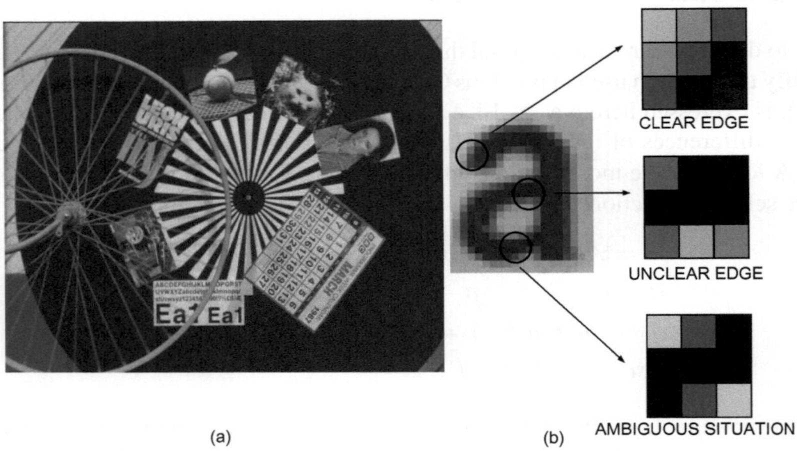

(a) (b)

Fig. 4.4. Examples of clear, unclear, and ambiguous edges for ELA algorithm.

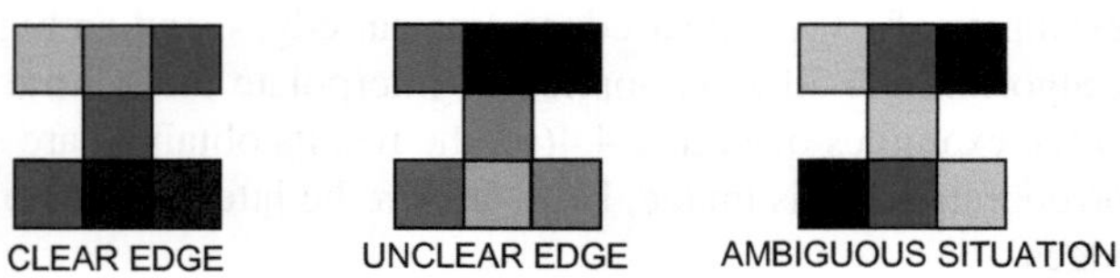

Fig. 4.5. ELA results for clear, unclear, and ambiguous edges.

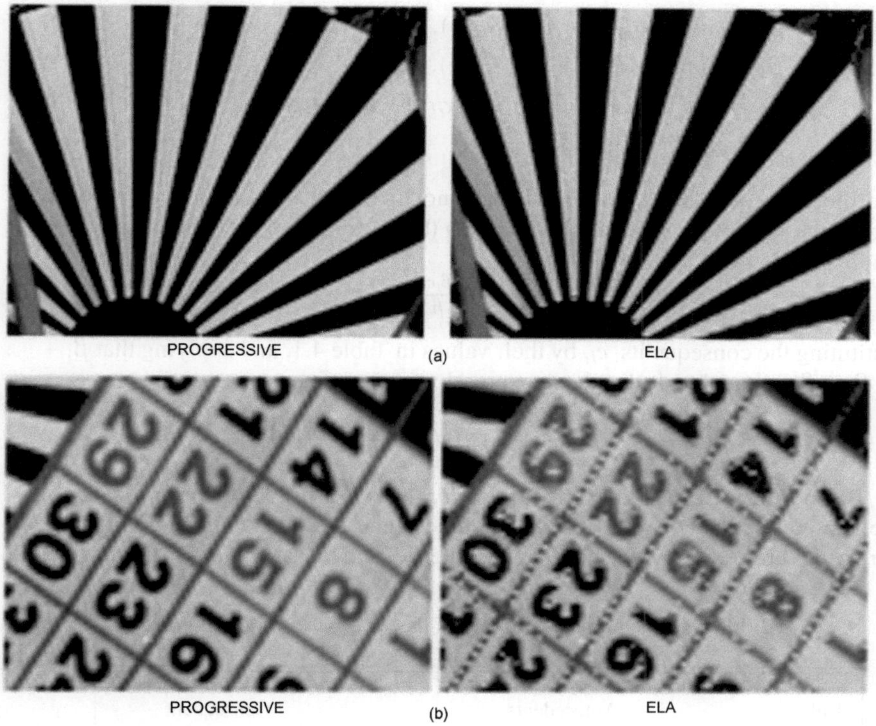

Fig. 4.6. Progressive images versus ELA results: (a) clear-defined edges, (b) unclear and ambiguous situations.

Our approach applies heuristic knowledge to overcome these limitations. The following knowledge is employed to estimate correctly the edge direction without enlarging the processing window:

1. An edge is clear in direction a not only if a is small but also if b and c are large.
2. An edge is clear in direction c not only if c is small but also if a and b are large.
3. If a and c are very small and b is large, neither there is an edge nor vertical linear interpolation performs well; the best option is a linear interpolation between the neighbors with small differences: A, C, D, F.
4. Otherwise, a vertical linear interpolation would be the most adequate.

This heuristic knowledge is fuzzy since the concepts of 'small', 'large', and 'very small' are not understood as threshold values but as fuzzy ones. Hence, our proposal is to model this knowledge by a fuzzy system. The rule base of the fuzzy system is described in Table 4.1. The concepts of *SMALL*, *LARGE* and *VERY_SMALL* are represented by fuzzy sets whose membership functions change continuously instead of abruptly between 0 and 1 membership values (μ), as shown in Figure 4.7.

Since the minimum operator is used as connective 'and' of antecedents, the activation degrees of the rules, β_i, are calculated as follows:

$$\beta_1 = min(\mu_{SMALL}(a), \mu_{LARGE}(b), \mu_{LARGE}(c)) \quad (4.2)$$
$$\beta_2 = min(\mu_{LARGE}(a), \mu_{LARGE}(b), \mu_{SMALL}(c)) \quad (4.3)$$
$$\beta_3 = min(\mu_{VERYSMALL}(a), \mu_{LARGE}(b), \mu_{VERYSMALL}(c)) \quad (4.4)$$
$$\beta_4 = 1 - \beta_1 - \beta_2 - \beta_3 \quad (4.5)$$

Since the consequents, c_i, of the rules are not fuzzy, the global conclusion provided by the system is calculated by applying the Fuzzy Mean (FM) defuzzification method as follows:

$$X = \frac{\sum_{i=1}^{4} \beta_i c_i}{\sum_{i=1}^{4} \beta_i} \quad (4.6)$$

Substituting the consequents, c_i, by their values in Table 4.1, and applying that $\beta_1 + \beta_2 + \beta_3 + \beta_4$ is equal to 1, the above expression can be given as:

$$X = \beta_1\left(\frac{A+F}{2}\right) + \beta_2\left(\frac{C+D}{2}\right) + \beta_3\left(\frac{A+F+C+D}{4}\right) + \beta_4\left(\frac{B+E}{2}\right) \quad (4.7)$$

The 'Basic Fuzzy-ELA' algorithm applies *line average* in the directions a or c if there is a clear edge (β_1 or β_2 takes the value 1 and the others are 0). Otherwise, several rules are active and the new pixel is interpolated by using all or several values

Table 4.1. The proposed rule set of the fuzzy-ELA algorithm

Rule	Antecedents	Consequent
1.	a is *SMALL* and b is *LARGE* and c is *LARGE*	$X = (A+F)/2$
2.	a is *LARGE* and b is *LARGE* and c is *SMALL*	$X = (C+D)/2$
3.	a is *VERY_SMALL* and b is *LARGE* and c is *VERY_SMALL*	$X = (A+F+C+D)/4$
4.	*Otherwise*	$X = (B+E)/2$

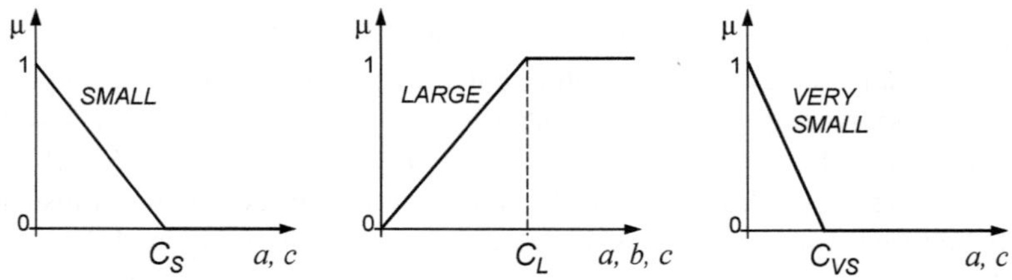

Fig. 4.7. Membership functions *SMALL*, *LARGE*, and *VERY_SMALL*.

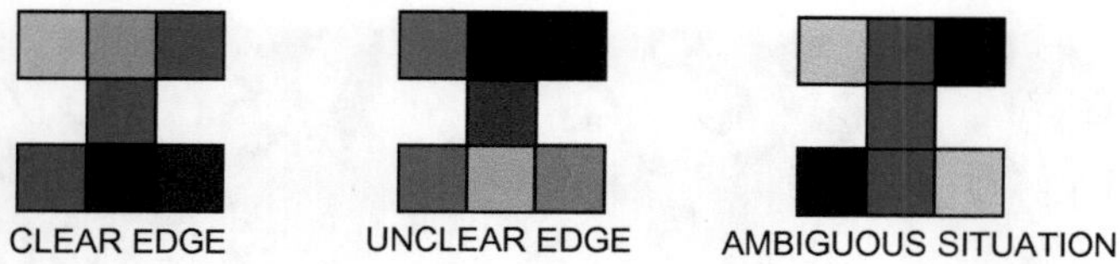

Fig. 4.8. 'Basic Fuzzy-ELA' results for clear, non-clear and ambiguous edges.

of the 3+3 pixels (A, B, C, D, E, F). As a consequence, the proposed method works as well as ELA with clear edges and better than ELA with unclear and ambiguous edges. This is illustrated in Figure 4.8, where the examples in Figure 4.4 have been processed with this 'Basic Fuzzy-ELA' algorithm.

Other examples to show up how the 'Basic Fuzzy-ELA' works are illustrated in Figures 4.9, 4.10, and 4.11. All of them are zooms of the de-interlaced image in Figure 4.1 after applying *line average*, ELA, and the 'Basic Fuzzy-ELA' algorithms. Figure 4.9 and 4.10 correspond to areas of the image where there are clearly defined edges. In both cases ELA obtains the best results since it reconstructs the edges more sharply while *line average* introduces an undesirable effect that produces a blurred perception of the image. The 'Basic Fuzzy-ELA' algorithm gets better results than

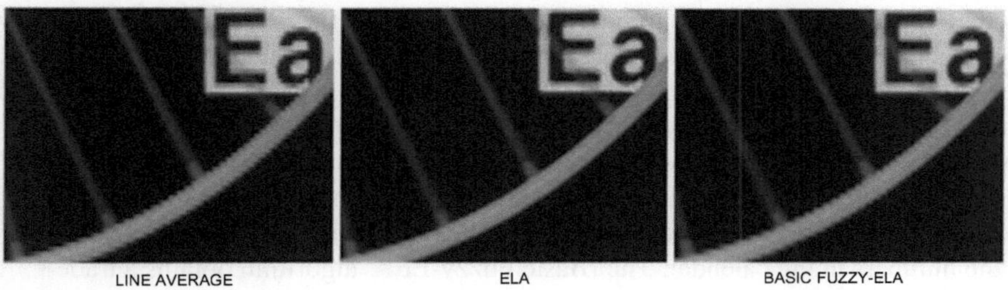

Fig. 4.9. Examples of clear edges in Figure 4.1. The corresponding images obtained after de-interlacing by *line average*, ELA, and the 'Basic Fuzzy-ELA' algorithm.

Fig. 4.10. Examples of an area with clear edges in Figure 4.1. The corresponding images obtained after de-interlacing by *line average*, ELA, and the 'Basic Fuzzy-ELA' algorithm.

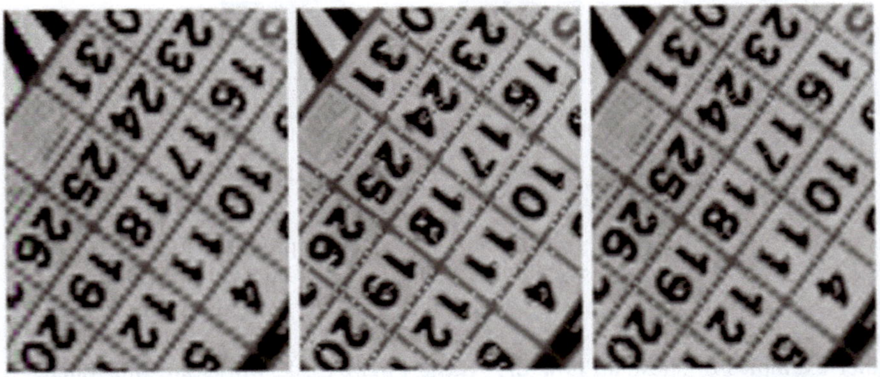

Fig. 4.11. Examples of an area with unclear edges in Figure 4.1. The corresponding images obtained after de-interlacing by *line average*, ELA, and the 'Basic Fuzzy-ELA' algorithm.

line average but its performance compared to ELA is inferior in the reconstruction of clear edges in both pictures. Therefore, the avoidance of ELA mistakes has a penalty in the lack of quality of clear edges. This seems logical since the fuzzy system demands more premises to interpolate according to the presence of an edge.

Figure 4.11 shows the results after de-interlacing a detailed area of the image in Figure 4.1 with *line average*, ELA, and 'Basic Fuzzy-ELA' algorithms. The performance of all these algorithms is poor since this area contains a high number of fine details that bring about the increase of ambiguous and unclear situations. As shown in Figure 4.11, *line average* offers a blurred de-interlaced image whereas ELA introduces serious mistakes in the reconstruction of the fine details, for instance the lines and the numbers in the calendar. The 'Basic Fuzzy-ELA' algorithm obtains a trade-off between the performance of both algorithms since it eliminates the majority of ELA mistakes and reduces the blur in some edges compared with *line average*.

4.1.1 Determination of the Membership Function Parameters

The rules of the proposed fuzzy algorithm have been defined from heuristic knowledge. However, heuristic knowledge does not provide much information about the definition of the membership functions. The shape of trapezoids (see Figure 4.7) was chosen since its implementation is easy, but there are many possibilities to select the parameters C_S, C_L, and C_{VS} that describe the membership functions of the concepts *SMALL*, *LARGE*, and *VERY_SMALL*. In fact, to make the concepts of *SMALL* and *VERY_SMALL* meaningful there is only a restriction: C_{VS} should be smaller than C_S. Another consideration is that the activation degree of rule 4 should not be negative, that is, $\beta_1 + \beta_2 + \beta_3$ should always be smaller or equal to one. Among the lot of values that meet these constraints, some of them will surely provide better results than others. If a set of data is available from progressive images, an error function can evaluate the difference between using some parameters or others, and supervised learning algorithms can be applied to minimize the error. This is the approach we have followed to select good values for the membership function parameters.

Automatic tuning has been applied by using the CAD tools integrated into the Xfuzzy 3.1 development environment. The fuzzy system employed in the 'Basic Fuzzy-ELA' algorithm has been described within Xfuzzy 3.1 as a first-order Takagi-Sugeno system, since the rule consequents can be considered as linear functions of the inputs. To meet the requirement of calculating the activation degree β_4 as $1-\beta_1-\beta_2-\beta_3$, the rule set in Table 4.1 has been translated into the equivalent one shown in Table 4.2, where the consequents are defined as follows:

Table 4.2. Description of the 'Basic Fuzzy-ELA' rule set with Xfuzzy 3.1

Rule	Antecedents	Consequent
1.	*a* is *SMALL* and *b* is *LARGE* and *c* is *LARGE*	c_1
2.	*a* is *LARGE* and *b* is *LARGE* and *c* is *SMALL*	c_2
3.	*a* is *VERY_SMALL* and *b* is *LARGE* and *c* is *VERY_SMALL*	c_3
4.	*Anycase*	c_4

$$c_1 = \frac{(A+F)}{2} - \frac{(B+E)}{2} \tag{4.8}$$

$$c_2 = \frac{(C+D)}{2} - \frac{(B+E)}{2} \tag{4.9}$$

$$c_3 = \frac{(A+C+D+F)}{4} - \frac{(B+E)}{2} \tag{4.10}$$

$$c_4 = \frac{(B+E)}{2} \tag{4.11}$$

The final conclusion of the fuzzy system implemented in Xfuzzy 3.1 is obtained after applying a weighted sum as defuzzication method:

$$\begin{aligned} X = {} & \beta_1\left(\tfrac{A+F}{2} - \tfrac{B+E}{2}\right) + \beta_2\left(\tfrac{C+D}{2} - \tfrac{B+E}{2}\right) \\ & +\beta_3\left(\tfrac{A+F+C+D}{4} - \tfrac{B+E}{2}\right) + \left(\tfrac{B+E}{2}\right) \end{aligned} \tag{4.12}$$

which is equivalent to the expression (4.7), as desired.

Figures 4.12 and 4.13 illustrate the graphical user interfaces of the CAD tool *xfedit* of Xfuzzy 3.1, which have been employed to describe this fuzzy system.

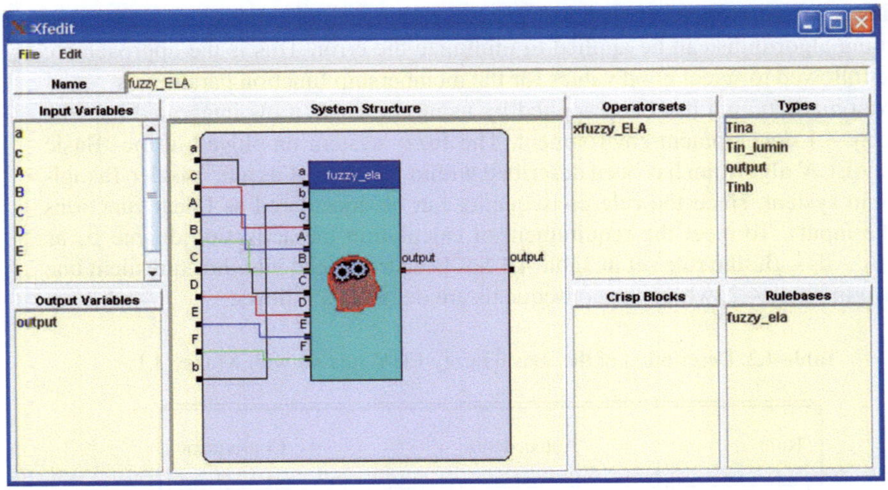

Fig. 4.12. The fuzzy system implemented in the Xfuzzy 3.1 environment.

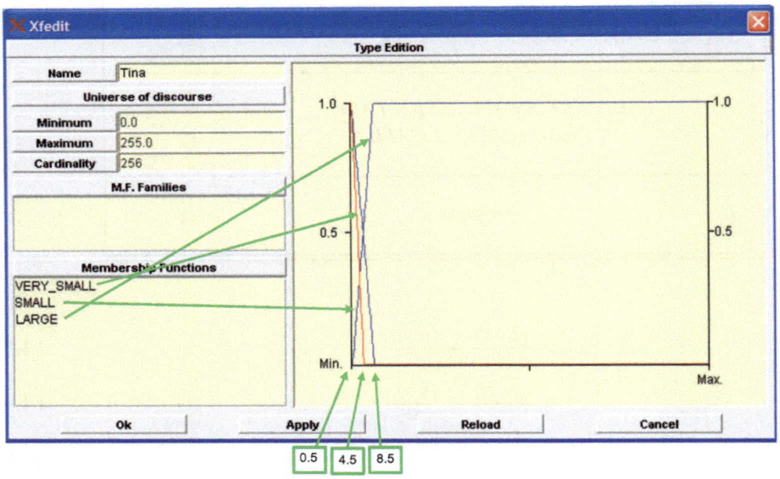

Fig. 4.13. Description of the membership functions in the Xfuzzy 3.1 environment.

Tuning of this fuzzy system by supervised learning has been done thanks to another CAD tool of Xfuzzy 3.1 named *xfsl*. Among the wide set of algorithms available within this tool, Marquardt-Levenberg algorithm has been employed. Exploiting that the *xfsl* tool allows the user to configure the tuning process, only the parameters involved in the description of the membership functions *SMALL*, *LARGE*, and *VERY_SMALL* have been enabled to participate in the tuning process, as shown in Figure 4.14. Because of the symmetry regarding the input variables 'a' and 'c', both share the same membership functions while the input variable 'b' has its own concept of *LARGE*.

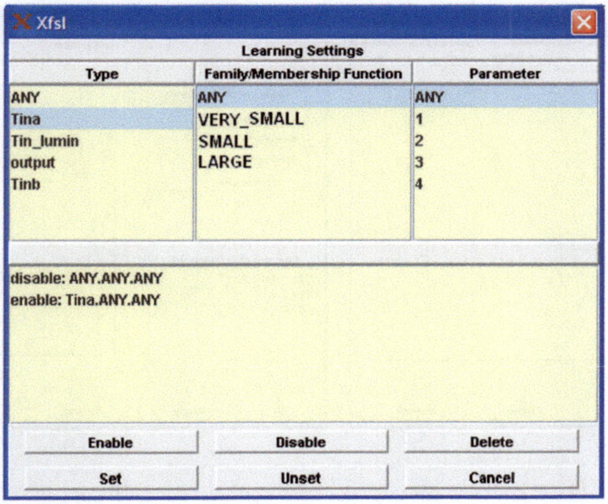

Fig. 4.14. Configuration of the tuning process.

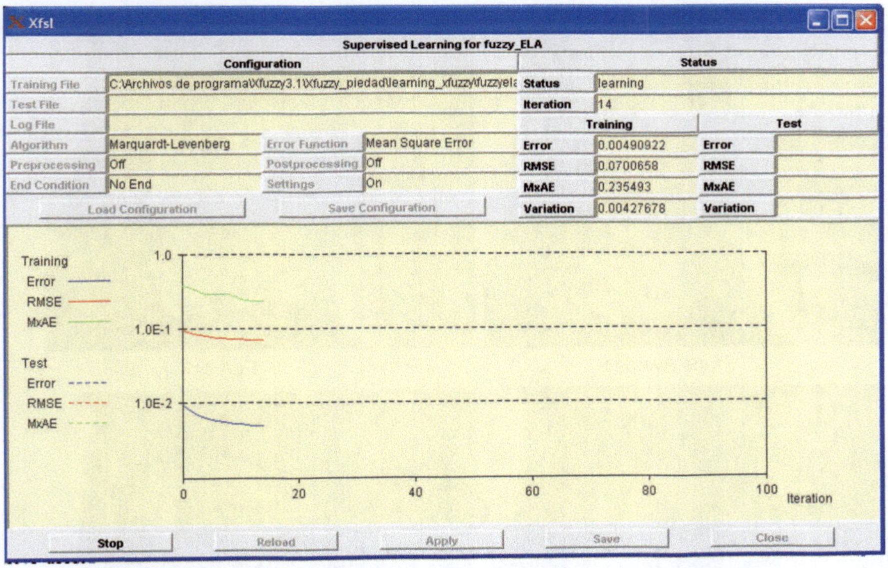

Fig. 4.15. Evolution of the tuning process.

The set of input/output training patterns have been generated by using 2 or 3 frames from all the video sequences presented in Chapter 2 so as to consider the highest number of different situations (from *VERY_SMALL* to *LARGE* luminance correlations). Figure 4.15 shows the evolution of the error functions during the learning process.

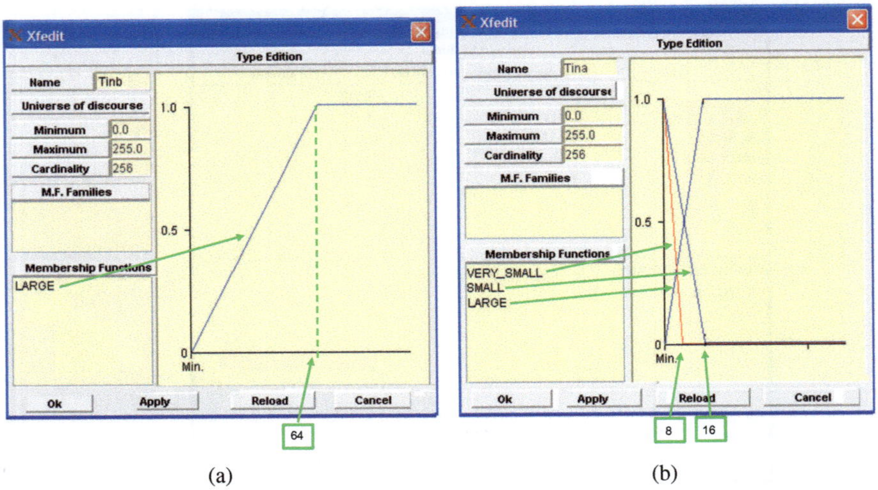

Fig. 4.16. Results after the learning process for (a) the input variable 'b' and (b) the input variables 'a' and 'c'.

Fig. 4.17. De-interlaced images for the Carphone sequence.

The result of the learning process is that the parameter C_{VS} of the *VERY_SMALL* membership function takes a value close to 8; the parameters C_S and C_L of *SMALL* and *LARGE* functions are close to 16 for the input variables 'a' and 'c', while C_L is approximately 64 for the input variable '*b*', as shown in Figure 4.16. This means that a significant variation has been introduced by the tuning process (compare Figures 4.13 and 4.16). The values of 64, 16 and 8 are also very efficient from a hardware point of view, and, hence, they have been chosen to complete the definition of the 'Basic Fuzzy-ELA' algorithm.

4.1.2 Performance of the Basic Fuzzy-ELA Algorithm

The performance of the 'Basic Fuzzy-ELA' algorithm has been proved by de-interlacing the sequences used in the previous Chapters. Table 4.3 shows the average value of PSNR after de-interlacing the sequences with QCIF, CIF, DVD, and film formats. Table 4.3 also includes the results obtained with *line average* and other edge-adaptive de-interlacing algorithms: conventional ELA, and the enhanced ELA algorithm in [4][1].

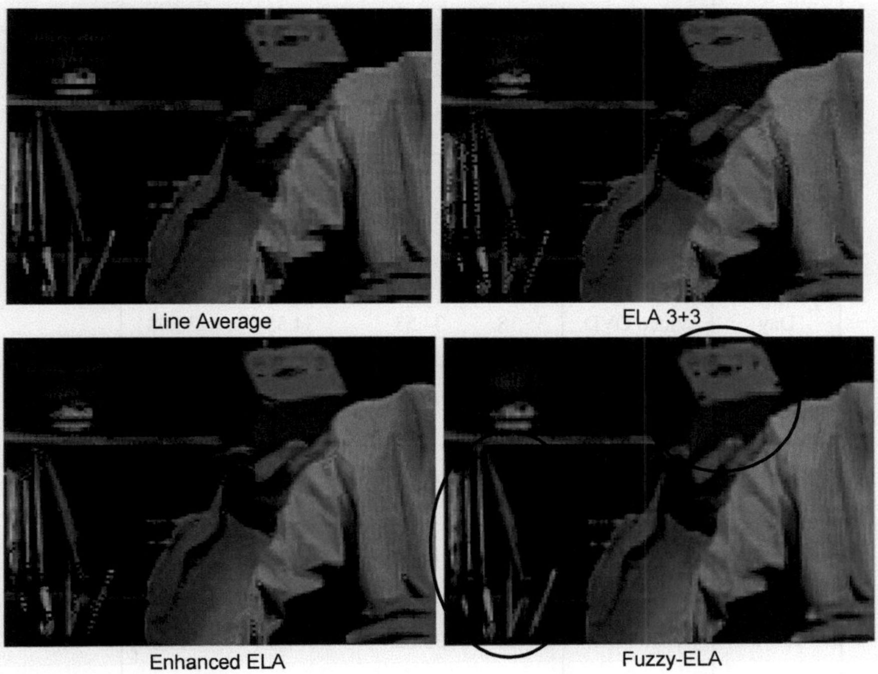

Fig. 4.18. De-interlaced images for the Salesman sequence.

[1] The algorithm presented in [4] consists of a modified ELA module and a contrast enhancement module. The results in Table 4.3 are obtained with an implementation of this algorithm that only uses the ELA module.

Table 4.3. PSNR results (in dBs) for different edge-adaptive methods

Sequence	Type of material	Line Average	ELA 3+3	Enhanced ELA [4]	Fuzzy-ELA
News	Video QCIF	29.25	26.63	27.22	28.91
Mother	Video QCIF	35.94	35.39	35.98	36.08
Carphone	Video QCIF	32.61	32.65	32.84	32.73
Missa	Video CIF	40.47	39.49	40.02	40.69
Paris	Video CIF	26.67	25.53	25.61	26.57
Salesman	Video CIF	35.04	32.11	32.35	33.42
Trevor	Video CIF	33.53	34.11	34.63	34.98
Phantom Menace	Video DVD	34.14	33.91	34.04	34.08
Die Another Day	Video DVD	35.18	34.53	34.55	35.08
Fire Rose	Film PAL	38.76	35.55	36.31	38.28
Fargo1	Film PAL	35.92	35.28	35.49	35.85
Fargo2	Film PAL	34.31	33.66	33.63	33.81
Tokyo	Film PAL	31.46	30.02	30.21	32.02
Bodyguard	Film PAL	35.61	35.08	34.93	35.94

Interesting data in Table 4.3 are that the simple *line average* method obtains the best PSNR results in several sequences (News, Paris, Salesman, etc.). The reason is that the main advantage of using edge-adaptive proposals is the better reconstruction of edges of the interpolated image and this does not always imply an important reduction of the overall error measurement. However, the visual inspection of the interpolated images makes evident the advantages introduced by edge-adaptive proposals. To illustrate this, several de-interlaced frames of the three sequences chosen in Chapter 2 (Carphone, Salesman, and Fargo2) are shown. Specially relevant are the results achieved in the Carphone sequence that has a high number of clear edges. Figure 4.17 shows the sixth frame of this sequence after applying the edge-adaptive algorithms included in Table 4.3. At naked eye, the results are clearly favorable for the edge-adaptive proposals over *line average*. The visual inspection of the Salesman frame (in Figure 4.18) and the Fargo2 frame (in Figure 4.19) is also advantageous to edge-adaptive proposals over *line average*.

In the other side, Table 4.3 shows that conventional ELA 3+3 and the enhanced ELA are almost always exceeded by the 'Basic Fuzzy-ELA' approach in terms of PSNR values. Visual inspection also corroborates this enhancement. In the Carphone frame, for instance, conventional ELA reconstructs quite well the edges in

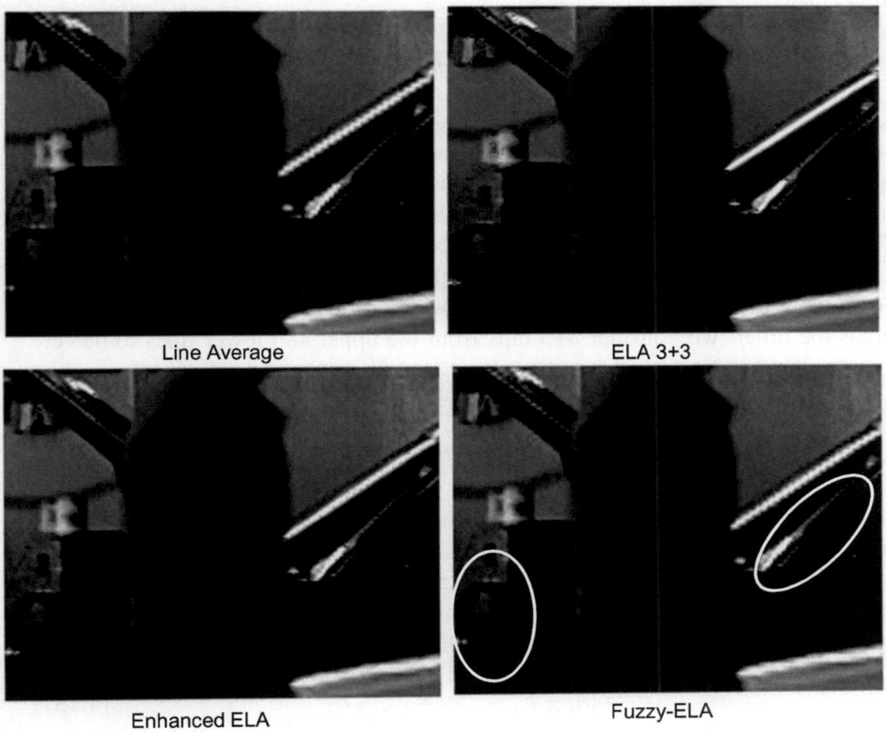

Fig. 4.19. De-interlaced images for the Fargo2 sequence.

the rear window but it introduces annoying mistakes, such as the black points in the man's ear and the frame of the window car. The enhanced ELA in [4] is not able to eliminate all these mistakes but reduces quite of them. The 'Basic Fuzzy-ELA' algorithm does not present the above commented errors but the quality of the resultant image in certain areas is worse. In particular, the reconstruction of edges is scarce around bow tie, the right man's shoulder, and the rear window of the car (marked with black ellipses).

Figure 4.18 illustrates that conventional edge-adaptive algorithms (ELA and enhanced ELA) introduce serious mistakes in the left part of the images in the Salesman sequence (pencils and books on the shelf). However, the edges of the box and man's sleeve are properly reconstructed in comparison with the blurred effect obtained with *line average*. The 'Basic Fuzzy-ELA' improves the results in the left part of the image, and also obtain good results in the reconstruction of the squared box and the salesman's shoulder (areas inside the black ellipses in Figure 4.18).

Figure 4.19 illustrates that the conventional ELA and enhanced ELA algorithms introduce serious mistakes in the reconstruction of some edges in the Fargo2 images (see black points in Figure 4.19). These errors are eliminated by the 'Basic Fuzzy-ELA' algorithm but at expense of losing some sharp edges such as those of the cars.

4.2 Modifications of the Basic Fuzzy-ELA Algorithm

Three extensions of the 'Basic Fuzzy-ELA' algorithm have been explored in order to improve its performance. Two strategies are studied: one is focused on ensuring the presence of an edge in areas with non-clear defined edges. The other increases the number of edge directions analyzed from three to five.

The new approaches also employ fuzzy inference systems to perform a non-linear interpolation, but they work with a larger neighborhood of pixels. The first one uses samples from the previously interpolated row to ensure the presence of an edge, whereas the other two consider 5+5 taps from the upper and lower lines to detect more edge directions.

4.2.1 Recursive Fuzzy-ELA Algorithm

This approach increases the consistency of the direction in which the interpolation is made. In particular, it ensures the presence of an edge in the a and c orientations, which correspond to the angles 45° and 135°, respectively (see Figure 4.20). For that purpose, the pixels A_i and C_i from the previously de-interlaced line are considered in two new inputs of this fuzzy system, as follows[2]:

$$a_i = |A_i - F| \qquad c_i = |C_i - D| \tag{4.13}$$

[2] To calculate the first line of the field, the 'Basic Fuzzy-ELA' algorithm is used.

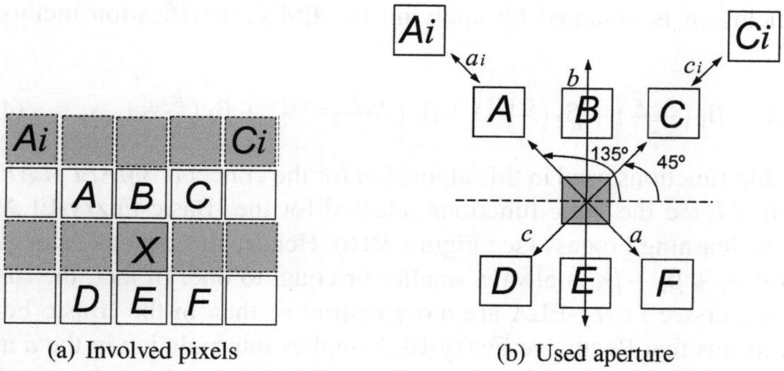

Fig. 4.20. Recursive Fuzzy-ELA algorithm.

Table 4.4. Description of the Recursive Fuzzy-ELA rule set

Rule	Antecedents	Consequent
1.	a is *SMALL* and a_i is *SMALL* and b is *LARGE* and c is *LARGE*	$X = \frac{A+F}{2}$
2.	a is *LARGE* and b is *LARGE* and c is *SMALL* and c_i is *SMALL*	$X = \frac{C+D}{2}$
3.	a is *VERY_SMALL* and b is *LARGE* and c is *VERY_SMALL*	$X = \frac{A+F+C+D}{4}$
4.	*Otherwise*	$X = \frac{B+E}{2}$

The algorithm implements the four rules described in Table 4.4. Compared with the rules of the 'Basic Fuzzy-ELA', the first and second rules have a higher number of antecedents (see Table 4.1). The activation degree of the rules are calculated by using the minimum operator as connective 'and':

$$\beta_1 = \min(\mu_{SMALL}(a), \mu_{SMALL}(a_i), \mu_{LARGE}(b), \mu_{LARGE}(c)) \tag{4.14}$$

$$\beta_2 = \min(\mu_{LARGE}(a), \mu_{LARGE}(b), \mu_{SMALL}(c), \mu_{SMALL}(c_i)) \tag{4.15}$$

$$\beta_3 = \min(\mu_{VERY\_SMALL}(a), \mu_{LARGE}(b), \mu_{VERY\_SMALL}(c)) \tag{4.16}$$

$$\beta_4 = 1 - \beta_1 - \beta_2 - \beta_3 \tag{4.17}$$

The final conclusion is obtained by applying the FM defuzzification method as follows:

$$X = \beta_1 \left(\tfrac{A+F}{2}\right) + \beta_2 \left(\tfrac{C+D}{2}\right) + \beta_3 \left(\tfrac{A+F+C+D}{4}\right) + \beta_4 \left(\tfrac{B+E}{2}\right) \tag{4.18}$$

The membership functions used in this algorithm for the concepts *SMALL*, *LARGE*, and *VERY_SMALL* are the same functions selected for the 'Basic Fuzzy-ELA' after applying the learning process (see Figure 4.16). Hence, β_4 never takes negative values because $\beta_1 + \beta_2 + \beta_3$ is always smaller or equal to one. In fact, the rules 1 and 2 in the Recursive Fuzzy-ELA are more restrictive than in the 'Basic Fuzzy-ELA', which means that Recursive Fuzzy-ELA applies interpolation in the a and c orientations less frequently than the 'Basic Fuzzy-ELA', just when edges in those orientations are more evident.

4.2.2 ELA 5+5 and Fuzzy-ELA 5+5 Algorithm

The ELA algorithm originally proposed uses 3+3 taps from the upper and lower lines. A simple expansion in order to increase the edge directions analyzed is to consider 5+5 taps, as shown in Figure 4.21. Hence, two new directions a' and c', which correspond to the angles 149.04^o and 30.96^o, respectively, can be analyzed. The pseudo-code of the ELA 5+5 algorithm is as follows:

$$a = |A - F| \quad b = |B - E| \quad c = |C - D| \quad c' = |C' - D'| \tag{4.19}$$

$$\begin{aligned} &if \;\; min(a,b,c,a',c') = a \rightarrow X = (A+F)/2 \\ &elseif \;\; min(a,b,c,a',c') = c \rightarrow X = (C+D)/2 \\ &elseif \;\; min(a,b,c,a',c') = a' \rightarrow X = (A'+F')/2 \\ &elseif \;\; min(a,b,c,a',c') = c' \rightarrow X = (C'+D')/2 \\ &else \;\; \rightarrow X = (B+E)/2 \end{aligned} \tag{4.20}$$

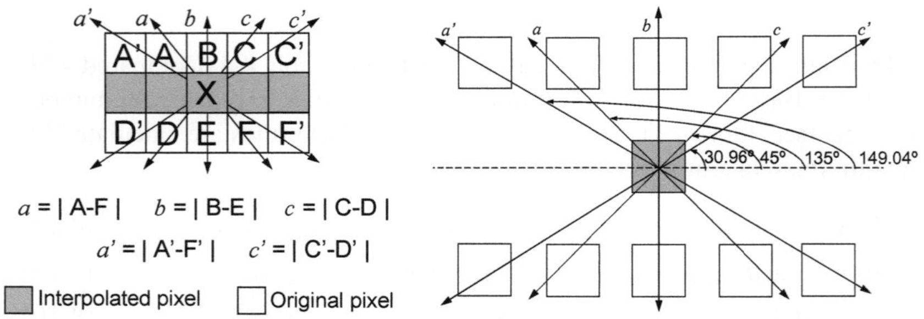

(a) 5+5 ELA scheme (b) 5+5 ELA aperture

Fig. 4.21. ELA algorithm with 5+5 pixels from the upper and lower lines.

Table 4.5. Description of the 'Fuzzy-ELA 5+5' rule set

Rule	Antecedents	Consequent
1.	a' is *LARGE* and a is *SMALL* and b is *LARGE* and c is *LARGE* and c' is *LARGE*	$X = \frac{A+F}{2}$
2.	a' is *LARGE* and a is *LARGE* and b is *LARGE* and c is *SMALL* and c' is *LARGE*	$X = \frac{C+D}{2}$
3.	a is *VERY_SMALL* and b is *LARGE* and c is *VERY_SMALL*	$X = \frac{A+F+C+D}{4}$
4.	a' is *SMALL* and a is *LARGE* and b is *LARGE* and c is *LARGE* and c' is *LARGE*	$X = \frac{A'+F'}{2}$
5.	a' is *LARGE* and a is *LARGE* and b is *LARGE* and c is *LARGE* and c' is *SMALL*	$X = \frac{C'+D'}{2}$
6.	*Otherwise*	$X = \frac{B+E}{2}$

Following the same philosophy used in the development of the 'Basic Fuzzy-ELA' algorithm, the ELA 5+5 could be improved by a new 'Fuzzy-ELA 5+5' approach. As a direct consequence of increasing the edge directions, the fuzzy inference system now applied extends the number of rules from 4 to 6. Furthermore, four of these rules (1, 2, 3, 4) have five instead of three antecedents as shown in Table 4.5. The concepts of *VERY_SMALL, SMALL, LARGE* used in Table 4.5 are again represented by the membership functions after the learning stage shown in Figure 4.16. The activation degrees of the rules, β_i, are calculated as follows:

$$\beta_1 = \mu_{LARGE}(a') \cdot \mu_{SMALL}(a) \cdot \mu_{LARGE}(b) \cdot \mu_{LARGE}(c) \cdot \mu_{LARGE}(c') \tag{4.21}$$
$$\beta_2 = \mu_{LARGE}(a') \cdot \mu_{LARGE}(a) \cdot \mu_{LARGE}(b) \cdot \mu_{SMALL}(c) \cdot \mu_{LARGE}(c') \tag{4.22}$$
$$\beta_3 = \mu_{VERY\_SMALL}(a) \cdot \mu_{LARGE}(b) \cdot \mu_{VERY\_SMALL}(c) \tag{4.23}$$
$$\beta_4 = \mu_{SMALL}(a') \cdot \mu_{LARGE}(a) \cdot \mu_{LARGE}(b) \cdot \mu_{LARGE}(c) \cdot \mu_{LARGE}(c') \tag{4.24}$$
$$\beta_5 = \mu_{LARGE}(a') \cdot \mu_{LARGE}(a) \cdot \mu_{LARGE}(b) \cdot \mu_{LARGE}(c) \cdot \mu_{SMALL}(c') \tag{4.25}$$
$$\beta_6 = 1 - \beta_1 - \beta_2 - \beta_3 - \beta_4 - \beta_5 \tag{4.26}$$

Although the product operator is more costly in hardware than the minimum operator, it is now used to represent the connective ‘and’ since this ensures that the activation degrees of the sixth rule never takes negative values.

The final conclusion is again obtained by using the FM defuzzification method as follows:

$$X = \beta_1\left(\frac{A'+F'}{2}\right) + \beta_2\left(\frac{C'+D'}{2}\right) + \beta_3\left(\frac{A+F}{2}\right) + \beta_4\left(\frac{C+D}{2}\right) + \beta_5\left(\frac{A+F+C+D}{4}\right) + \beta_6\left(\frac{B+E}{2}\right) \quad (4.27)$$

4.2.3 *Improved Fuzzy-ELA 5+5 Algorithm*

This algorithm also searches for edges in five directions (a', a, b, c, c'), but it uses a different strategy to achieve it. The idea is to include two more rules without modifying the rule base of the ‘Basic Fuzzy-ELA’ approach. Let us explain the new approach with Figure 4.22. One way of improving the rule base of the ‘Basic Fuzzy-ELA’ (shown in the left side of Figure 4.22) is to study the cases that are included in the fourth rule with the antecedent ‘*Otherwise*’. One of them is that the three antecedents are *LARGE*, that is, no edge is found in the three directions (a, b, and c). In this case, two new directions (a', c') are worth exploring (see the right side of Figure 4.22). The idea is to look for edges in the new directions only when the fourth rule of the ‘Basic Fuzzy-ELA’ is quite activated or partially activated.

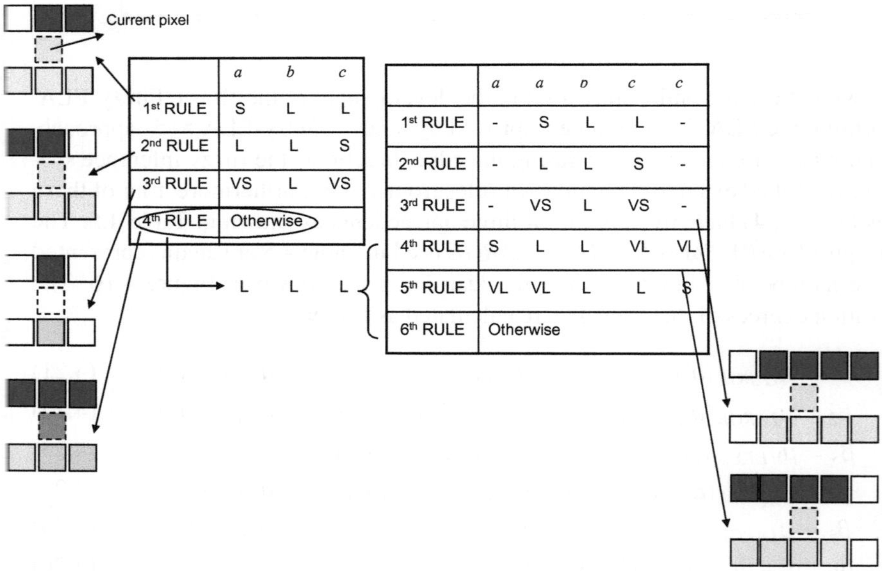

Fig. 4.22. Extension of the rule base in the ‘Improved Fuzzy-ELA 5+5’ algorithm.

Table 4.6. Description of the 'Improved Fuzzy-ELA 5+5' rule set

Rule	Antecedents	Consequent
1.	a is $SMALL$ and b is $LARGE$ and c is $LARGE$	$X = \frac{A+F}{2}$
2.	a is $LARGE$ and b is $LARGE$ and c is $SMALL$	$X = \frac{C+D}{2}$
3.	a is $VERY\_SMALL$ and b is $LARGE$ and c is $VERY\_SMALL$	$X = \frac{A+F+C+D}{4}$
4.	a' is $SMALL$ and a is $LARGE$ and b is $LARGE$ and c is $VERY\_LARGE$ and c' is $VERY\_LARGE$	$X = \frac{A'+F'}{2}$
5.	a' is $VERY\_LARGE$ and a is $VERY\_LARGE$ and b is $LARGE$ and c is $LARGE$ and c' is $SMALL$	$X = \frac{C'+D'}{2}$
6.	*Otherwise*	$X = \frac{B+E}{2}$

Table 4.6 shows the complete fuzzy rule set used by this algorithm. The fuzzy concepts are modeled by the membership functions shown in Figure 4.23 (the new rules, 4 and 5, include a new fuzzy concept called '$VERY\_LARGE$'). The overlapping of the membership functions has been selected to ensure that no more than two rules are simultaneously activated. This strategy allows the use of the minimum operator as connective 'and' since a positive value for the activation degree of the sixth rule is always obtained. The particular breakpoints of the membership functions (4, 20, 52, and 68 in Figure 4.23 have been obtained after applying a learning stage.

The activation degrees of the rules are as follows:

$$\beta_1 = min(\mu_{SMALL}(a), \mu_{LARGE}(b), \mu_{LARGE}(c)) \quad (4.28)$$

$$\beta_2 = min(\mu_{LARGE}(a), \mu_{LARGE}(b), \mu_{SMALL}(c)) \quad (4.29)$$

$$\beta_3 = min(\mu_{VERY\_SMALL}(a), \mu_{LARGE}(b), \mu_{VERY\_SMALL}(c)) \quad (4.30)$$

$$\beta_4 = min(\mu_{SMALL}(a'), \mu_{LARGE}(a), \mu_{LARGE}(b), \quad (4.31)$$

$$\mu_{VERY\_LARGE}(c), \mu_{VERY\_LARGE}(c')) \quad (4.32)$$

$$\beta_5 = min(\mu_{VERY\_LARGE}(a'), \mu_{VERY\_LARGE}(a), \mu_{LARGE}(b),$$

$$\mu_{LARGE}(c), \mu_{SMALL}(c')) \quad (4.33)$$

$$\beta_6 = 1 - \beta_1 - \beta_2 - \beta_3 - \beta_4 - \beta_5 \quad (4.34)$$

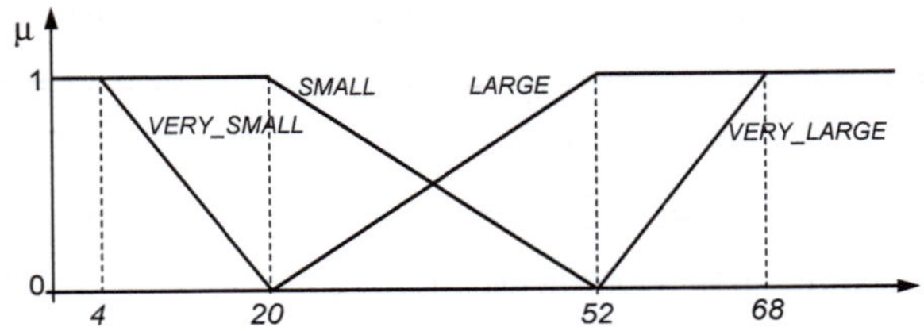

Fig. 4.23. Membership functions used by the Improved 5+5 Fuzzy-ELA algorithm.

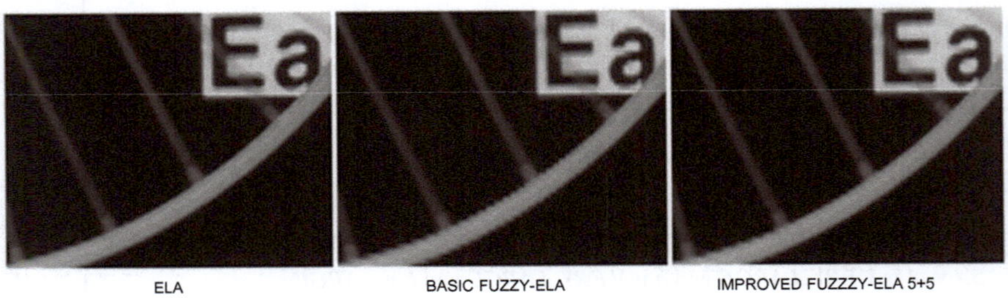

Fig. 4.24. Examples of clear edges in Figure 4.1.

The output provided by the algorithm (applying the FM defuzzication method) is as follows:

$$X = \beta_1 \left(\frac{A+F}{2}\right) + \beta_2 \left(\frac{C+D}{2}\right) + \beta_3 \left(\frac{A+F+C+D}{4}\right) + \beta_4 \left(\frac{A'+F'}{2}\right)$$
$$+\beta_5 \left(\frac{C'+D'}{4}\right) + \beta_6 \left(\frac{B+E}{2}\right) \quad (4.35)$$

The first simulation results obtained with this new algorithm offered few improvements in edges with degree inclinations in the directions a' and c' edges. One of the frequent mistakes was that the new two rules were frequently activated in cases of unclear edges. As result, the overall image quality was not as good as expected. To ensure the activation of the new rules in situations where an edge is really placed in a' and c' directions, the following considerations should be fulfilled:

- An edge is not an isolated feature of the image, that is, an edge does not belong to an unique pixel in the image.
- Since a and c directions are close to a' and c', it seems logical that the rules to interpolate along these edge directions (the first and fourth rules for a and a', and the second and fifth rules for c and c') should frequently be activated simultaneously.

Fig. 4.25. Examples of clear edges in Figure 4.1.

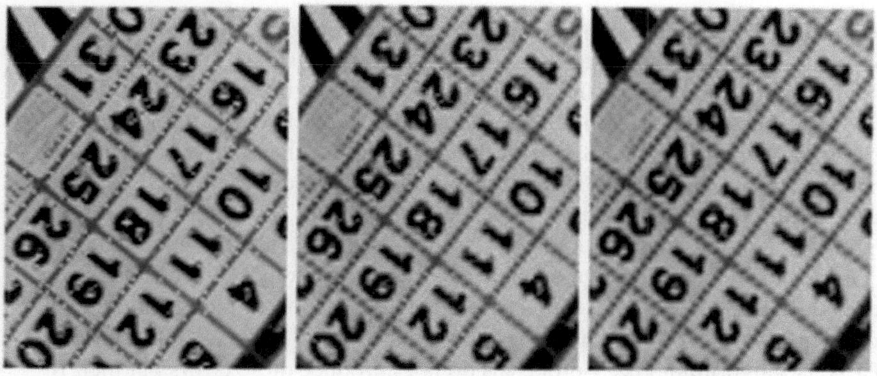

Fig. 4.26. Examples of an area with unclear edges in Figure 4.1.

To implement the above considerations, the following condition is imposed on the fourth rule:

$$\begin{aligned}
&if\ \ ((max(\beta_4|_{current}, \beta_1|_{current}) > 0)\ and\ (max(\beta_4|_{previous}, \beta_1|_{previous}) > 0)\ and\\
&(max(\beta_4|_{current}, \beta_1|_{current}) > (max(\beta_5|_{current}, \beta_2|_{current}))))\\
&then\ \ (\beta_4|_{current} = \beta_4|_{current}) \qquad (4.36)\\
&else\ \ (\beta_4|_{current} = 0)
\end{aligned}$$

where '*current*' means the current spatio-temporal coordinates of the pixel (x,y,t) and '*previous*' means the spatio-temporal coordinates of the previous pixel $(x-1,y,t)$.

And the corresponding condition for the fifth rule is as follows:

$$\begin{aligned} &if\ \ ((max(\beta_5|_{current}, \beta_2|_{current}) > 0)\ and\ (max(\beta_5|_{previous}, \beta_2|_{previous}) > 0)\ and \\ &(max(\beta_5|_{current}, \beta_2|_{current}) > (max(\beta_4|_{current}, \beta_1|_{current})))) \\ &then\ \ (\beta_5|_{current} = \beta_5|_{current}) \\ &else\ \ (\beta_5|_{current} = 0) \end{aligned} \tag{4.37}$$

The performance of this new fuzzy logic-based algorithm is qualitatively compared with the ELA and the 'Basic Fuzzy-ELA' algorithms in Figures 4.24, 4.25, and 4.26. As shown in these figures, the 'Improved Fuzzy-ELA 5+5' algorithm works better than the 'Basic Fuzzy-ELA' algorithm in clear and unclear edges.

4.2.4 Comparison of the Fuzzy-ELA Algorithms

Performance of the three modifications of the 'Basic Fuzzy-ELA' have been analyzed by de-interlacing the battery of sequences. Table 4.7 shows the average value of PSNR for the sequences with QCIF, CIF, and DVD formats. Table 4.7 also includes the results obtained by the 'Basic Fuzzy-ELA' and ELA algorithms with 3+3 and 5+5 taps. Comparing the four fuzzy-ELA algorithms, the highest PSNR values are obtained by the 'Improved Fuzzy-ELA 5+5'. It also achieves better PSNR values than ELA with 3+3 and 5+5 taps.

Figure 4.27 shows a snapshot of the Carphone sequence after de-interlacing by using ELA and all the fuzzy edge-adaptive algorithms. Although ELA 5+5 performs better than ELA 3+3 by visual inspection (see the reduction of the black points in the car window), ELA 5+5 is not able to eliminate all the mistakes introduced by the ELA 3+3 algorithm. Fuzzy proposals do not introduce these errors but their results in clear edges are poorer except for the 'Improved Fuzzy-ELA 5+5' algorithm, which provides the best performance (see the areas marked with black ellipses in Figure 4.27).

Figures 4.28 and 4.29 show the images obtained for one snapshot of a sequence called 'Flag'. This video sequence contains a high number of edges that make evident the efficiency of the 'Improved Fuzzy-ELA 5+5' algorithm versus *line average* (see Figure 4.28). The superior performance of the edge-adaptive approach is noticed at naked eye. The comparison among the edge-adaptive proposals is established in Figure 4.29. A zoom of a detailed area with unclear edges reveals ELA mistakes (white and black pixels). The increase of the ELA processing window up to 5+5 taps introduces more errors.The Basic, Recursive, and 5+5 Fuzzy-ELA algorithms reduce the errors at expense of a loss in the reconstruction of stripped lines of the flag. Finally, the 'Improved Fuzzy-ELA 5+5' algorithm obtains the best performance with a good reconstruction of clear edges and a considerable reduction of the errors.

A visual comparison of images from the Fargo2 sequence can be carried out in Figure 4.30. It corroborates a reduction of ELA 3+3 mistakes when ELA 5+5 is used (black points in the car). Although the fuzzy approaches eliminate all these

Fig. 4.27. De-interlaced images for the Carphone sequence.

errors, their performance are worse in the clear-defined edges of the car except for the 'Improved Fuzzy-ELA 5+5' algorithm, which obtains a better reconstruction of the clear edges as shown inside the black circumference.

Table 4.8 shows the resources used in the implementation of the four fuzzy algorithms. Since all of the algorithms described in this Chapter are spatial algorithms, their requirements in terms of storage elements are scarce. They do not require the inclusion of field buffers to store a complete field, and only line buffers

Table 4.7. PSNR results (in dBs) for different edge-adaptive methods

Sequence	Type of material	ELA 3+3	ELA 5+5	Basic Fuzzy-ELA	Recursive Fuzzy-ELA	Fuzzy-ELA 5+5	Improved Fuzzy-ELA 5+5
News	Video QCIF	26.63	25.92	28.91	27.55	28.17	29.21
Mother	Video QCIF	35.39	34.2	36.08	35.98	35.92	36.21
Carphone	Video QCIF	32.65	31.51	32.73	32.66	32.66	32.92
Missa	Video CIF	39.49	38.56	40.69	40.37	40.45	40.71
Paris	Video CIF	25.53	24.64	26.57	26.24	26.54	26.66
Salesman	Video CIF	32.11	30.17	33.42	33.68	33.07	33.52
Trevor	Video CIF	34.11	33.31	34.98	34.94	34.96	35.24
Phantom Menace	Video DVD	33.91	33.11	34.08	34.06	34.03	34.09
Die Another Day	Video DVD	34.53	33.56	35.08	34.87	35.01	35.12
Fire Rose	Film PAL	35.55	33.61	38.28	38.26	38.29	38.57
Fargo1	Film PAL	35.28	34.33	35.85	35.63	35.76	35.89
Fargo2	Film PAL	33.66	32.16	33.81	34.07	34.08	34.34
Tokyo	Film PAL	30.02	28.53	32.02	31.42	31.81	32.19
Bodyguard	Film PAL	35.08	33.93	35.49	35.07	35.39	35.26

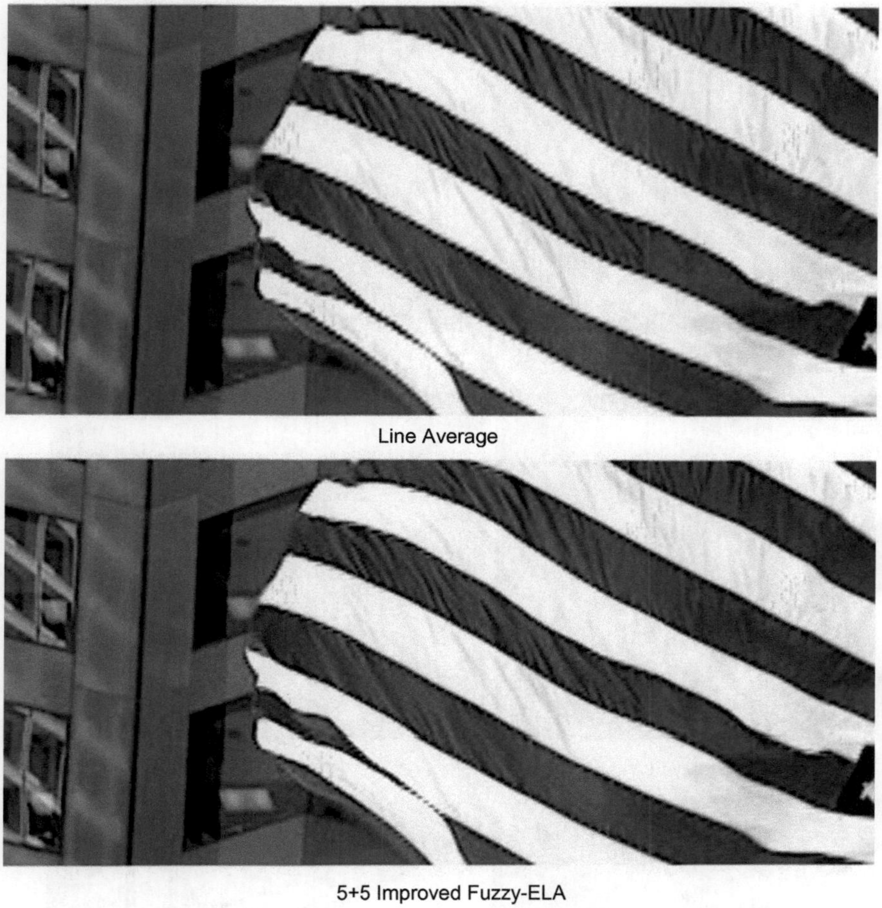

Fig. 4.28. De-interlaced images for the Flag sequence.

and registers are necessary to implement them. The basic proposal obviously requires the lowest number of resources. The increase of taps from 3+3 to 5+5 in the processing window implies two more registers in the 'Fuzzy-ELA 5+5' and 'Improved Fuzzy-ELA 5+5' algorithms. The 'Recursive Fuzzy-ELA' approach demands a second line buffer to provide the pixel values A_i and D_i (see Figure 4.20).

With regard to the number of primitive operations, the 'Basic Fuzzy-ELA' halves them in comparison with the rest of algorithms. However, none of the algorithms requires an excessive number since all of them use fuzzy logic-based systems with simple rules.

After examining the performance of all the fuzzy-ELA algorithms, we conclude that the 'Basic Fuzzy-ELA' is a good solution when looking for low computational

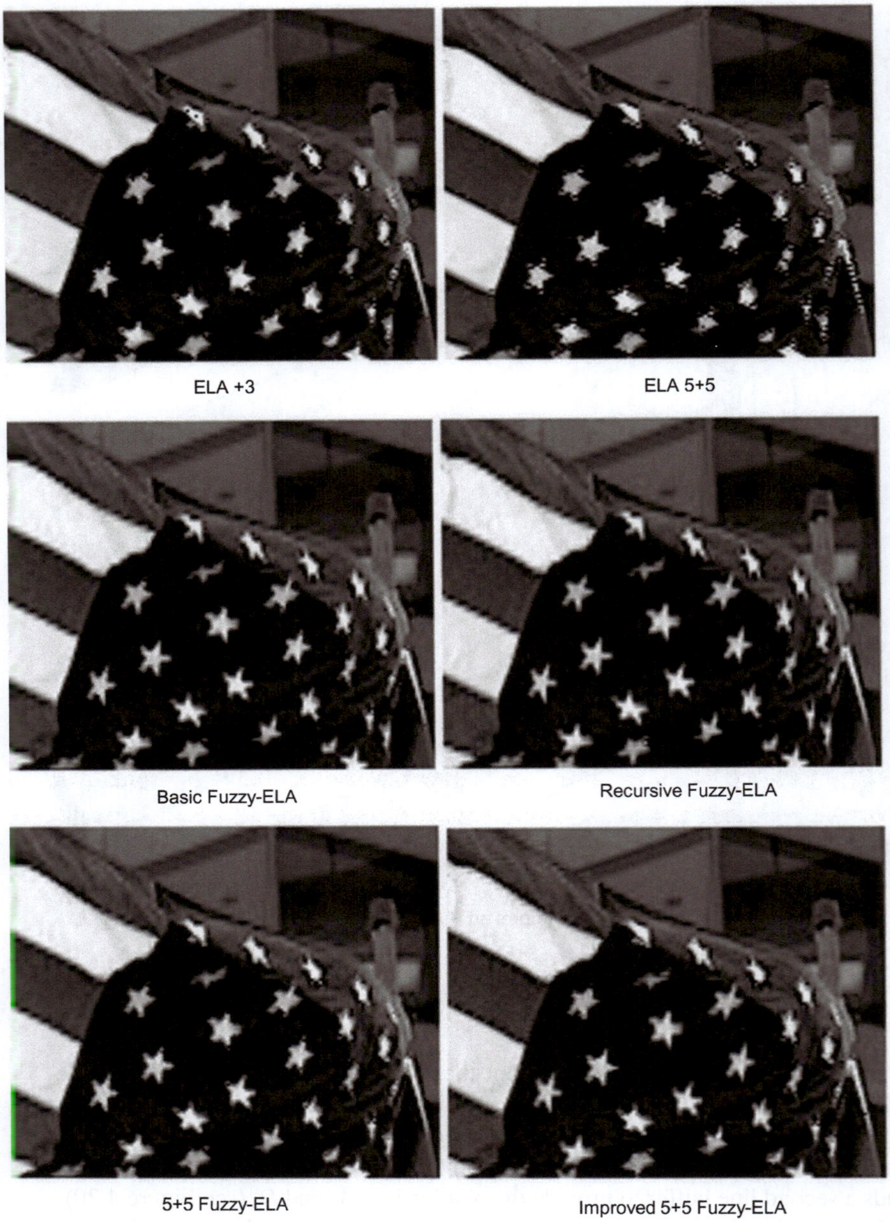

Fig. 4.29. A zoom of a de-interlaced snapshot for the Flag sequence.

cost. If a slight increase of computational cost is affordable, the 'Improved Fuzzy-ELA 5+5' is the best solution for its better reconstruction of edges. Furthermore, both of them surpass other spatial de-interlacing algorithms.

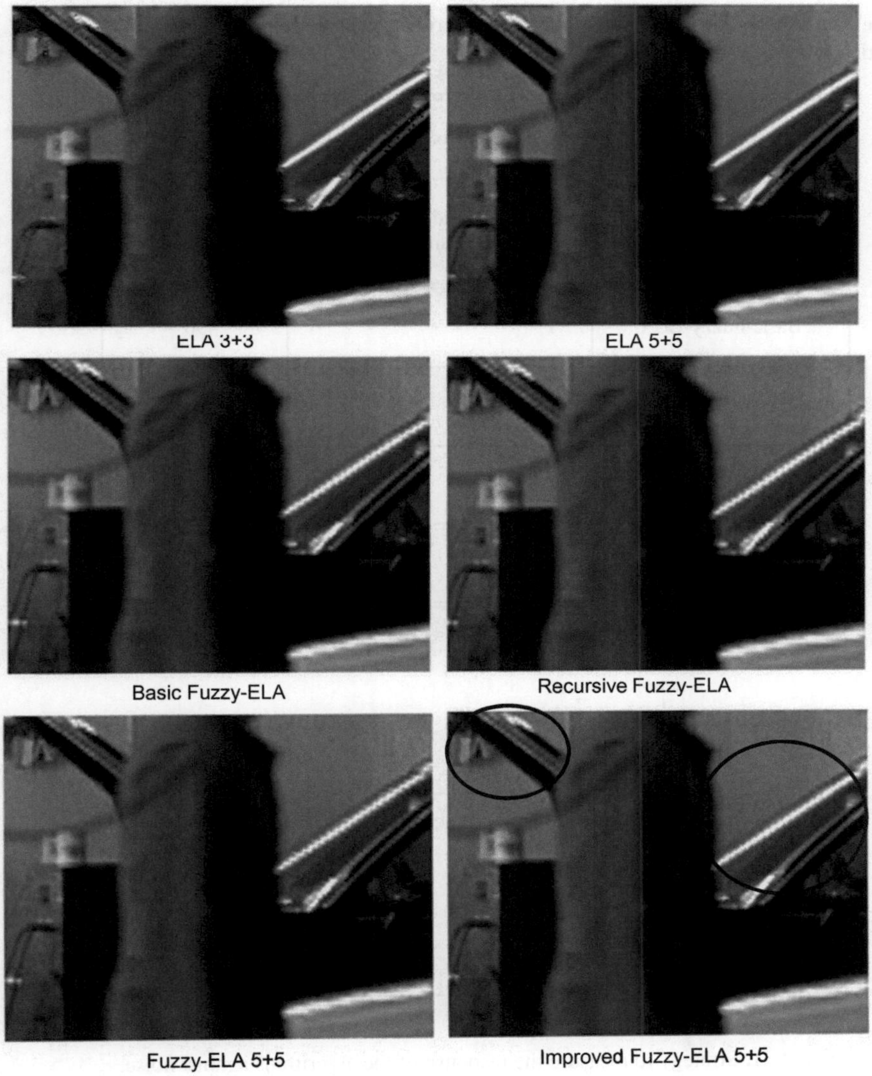

Fig. 4.30. De-interlaced images for the Fargo2 sequence.

4.2.5 Robustness of Fuzzy Proposals against Noise

The robustness of the proposed fuzzy edge-adaptive algorithms against noise has been explored by adding artificial noise to original sequences. Two kinds of noise are added: gaussian and impulse. Figure 4.31 shows the strategy used to add noise.

In case of impulse noise, a parameter called noise density controls the level of noise. This kind of noise consists of random occurrences of on and off pixels, that is,

Table 4.8. Analysis of storage resources and primitive operations (POs) of fuzzy edge-adaptive proposals

	Resources			Complexity
Algorithm	**Line buffer**	**Field memory**	**Register**	**Primitive Operations (POs)**
Basic Fuzzy-ELA	1	-	2	50
Recursive Fuzzy-ELA	2	-	4	98
Fuzzy-ELA 5+5	1	-	4	103
Improved Fuzzy-ELA 5+5	1	-	4	114

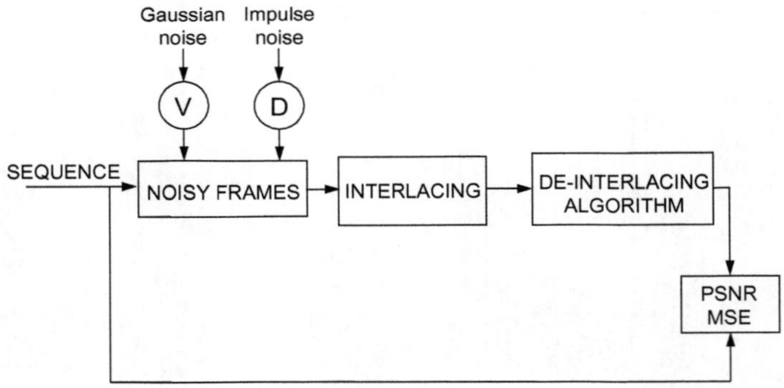

Fig. 4.31. Strategy to measure the response of the algorithms against noise.

white and black pixels. Noise density fixes the number (P) of total pixels in the frame affected by this type of noise. They are calculated by the following expression:

$$P = \frac{D \cdot M \cdot N}{100} \tag{4.38}$$

where D is the noise density in percentage and $M \cdot N$ is the resolution of a frame in the sequence.

Table 4.9 shows the results obtained by using several values of noise density in the Carphone sequence. For low values of noise density (1% and 4%), conventional

Table 4.9. PSNR results (in dBs) for the Carphone sequence with impulse noise

	D=1%	D=4%	D=10%	D=14%	D=16%
Line Average	25.46	20.06	16.23	14.77	14.22
ELA 3+3	26.82	21.46	17.32	15.69	15.01
Basic Fuzzy-ELA	26.45	21.16	17.19	15.63	15.05
Improved Fuzzy ELA 5+5	26.61	21.21	17.37	15.82	15.19

Table 4.10. PSNR results (in dBs) for the Carphone sequence with Gaussian noise

	V=0.002	V=0.004	V=0.006	V=0.008
Line Average	26.87	24.62	23.06	21.99
ELA 3+3	26.78	24.44	22.93	21.86
Basic Fuzzy-ELA	27.18	24.79	23.31	22.21
Improved Fuzzy ELA 5+5	27.04	24.74	23.29	22.22

ELA is able to avoid impulse noise and performs better than fuzzy edge-adaptive proposals. However, fuzzy proposals gradually reduces these differences in accordance with the increase of noise density. *Line average* obtains poor results by loosing more than 1 dB in PSNR values. In Figure 4.32, one snapshot of the Carphone sequence is shown after de-interlacing the noisy sequence with a noise density of 14%.

If the option of Gaussian noise is selected, then Gaussian white noise of variance (V) is added to the image. Table 4.10 shows the results obtained for the Carphone sequence using the following values of variance: 0.002, 0.004, 0.006, and 0.008.

The highest values of PSNR are obtained by the fuzzy edge-adaptive proposals. Their results are superior to *line average* and ELA algorithms as degradation of the image is higher. The performance of 'Basic Fuzzy-ELA' and the 'Improved Fuzzy-ELA 5+5' algorithm is very similar. ELA 3+3 does not improve the results

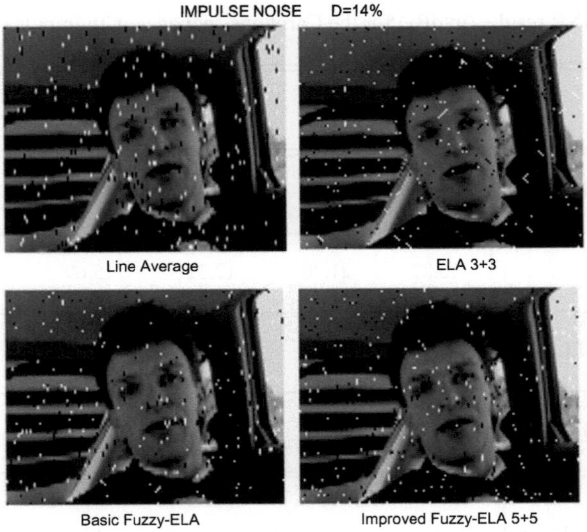

Fig. 4.32. De-interlaced images for the Carphone sequence with impulse noise.

Fig. 4.33. De-interlaced images for the Carphone sequence with Gaussian noise.

of fuzzy approaches in none of the cases of study. In Figure 4.33, one snapshot of the Carphone sequence with gaussian noise (V equals to 0.002) is shown after de-interlacing.

After de-interlacing the noisy sequences we can conclude that fuzzy approaches perform better than *line average* and ELA 3+3 algorithms. The advantages introduced are more evident as the noise degree increases.

Table 4.11. PSNR results (in dBs) for the proposals in Chapter 3 and Chapter 4

	News	Mother	Car-phone	Missa	Paris	Sales-man	Trevor	Phantom Menace	Die AnotherDay
	QCIF	QCIF	QCIF	CIF	CIF	CIF	CIF	DVD PAL	DVD NTSC
Proposal Chapter 3	38.28	41.35	35.05	40.51	35.77	38.45	37.29	34.39	42.35
Proposal Chapter 4	38.25	41.36	35.27	40.54	35.68	38.49	37.48	34.43	42.38

Fig. 4.34. De-interlaced images by the proposals in Chapter 3 (left) and Chapter 4 (right).

4.2.6 *Fuzzy Motion Adaptive Algorithm with the 'Improved Fuzzy-ELA 5+5' as Spatial Interpolator*

The comparative study of the fuzzy approaches with other conventional spatial de-interlacing algorithms has shown small improvements in PSNR values (less than

Fig. 4.35. A zoom of a de-interlaced frame in the Phantom Menace sequence (video material).

Table 4.12. PSNR results (in dBs) for the proposals in Chapter 3 and Chapter 4

	Fire Rose	Fargo1	Fargo2	Tokyo	Bodyguard
	PAL TV	PAL TV	PAL TV	PAL TV	PAL TV
Proposal Chapter 3	39.91	36.74	42.36	38.18	40.79
Proposal Chapter 4	39.99	36.65	42.29	38.21	40.61

Fig. 4.36. A zoom of a detailed area after de-interlacing (video material).

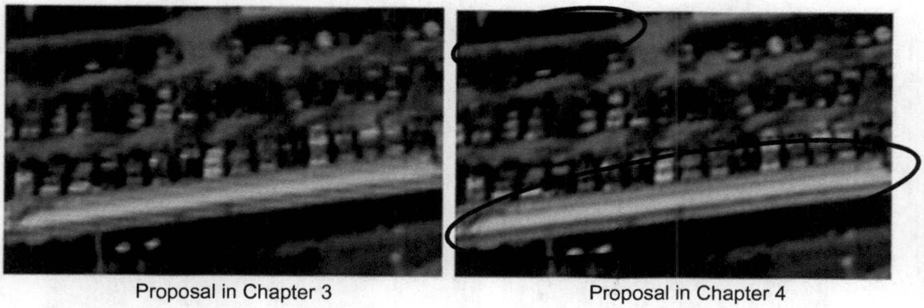

Fig. 4.37. A zoom of a de-interlaced frame in the Tokyo sequence (film material).

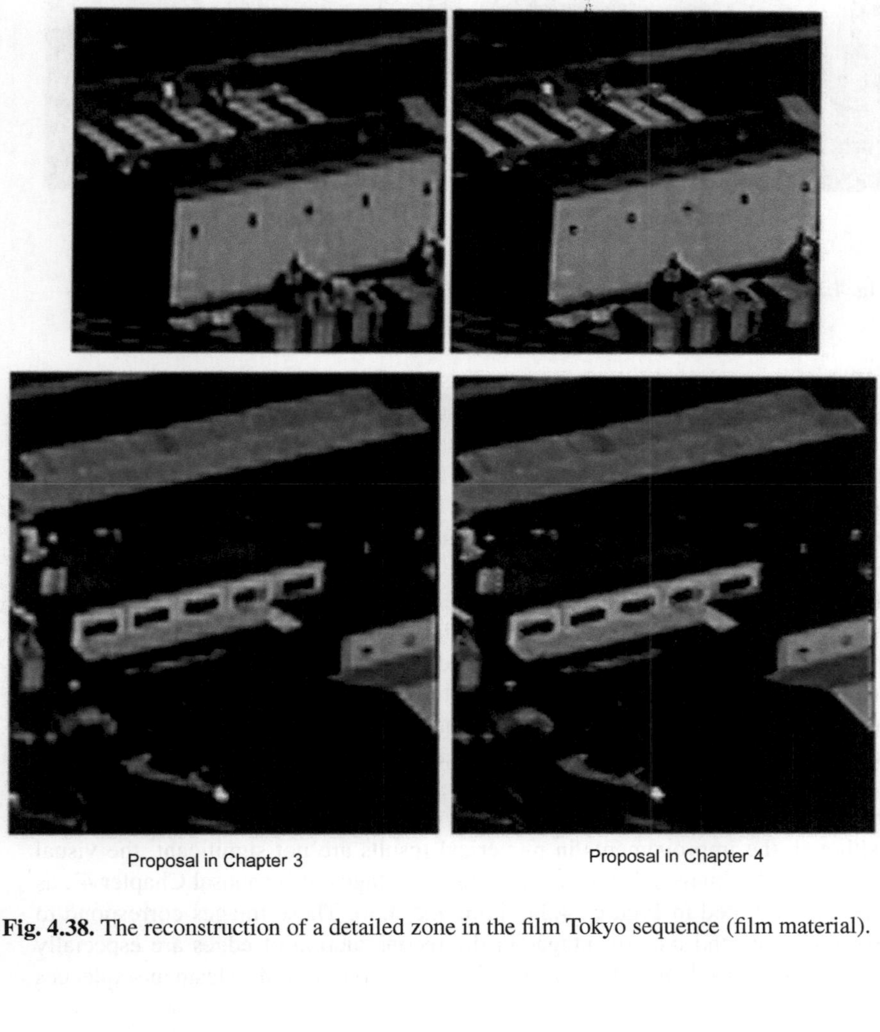

Fig. 4.38. The reconstruction of a detailed zone in the film Tokyo sequence (film material).

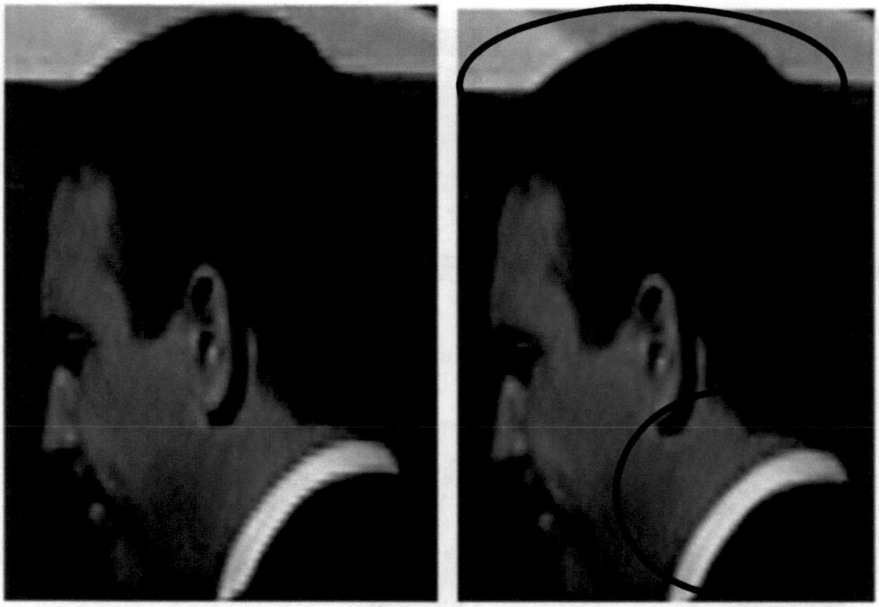

Fig. 4.39. A zoom of a de-interlaced frame in the Bodyguard sequence (film material).

0.5 dBs), but qualitatively, the improvements are clearly appreciated in areas of the images with edges. Since human visual system pays more attention to areas with edges in the image, a substantial improvement of them implies a positive perception of the viewer. Therefore, the use of the 'Improved Fuzzy-ELA 5+5' algorithm instead of *line average* in the fuzzy motion adaptive de-interlacing algorithm presented in Chapter 3 could introduce important improvements in the reconstruction of moving areas.

Tables 4.11 and 4.12 compare the fuzzy motion-adaptive algorithm selected in Chapter 3, which uses *line average* as spatial interpolator ('proposal Chapter 3'), with the same algorithm but using 'Improved Fuzzy-ELA 5+5' as spatial interpolator ('proposal Chapter 4'). After analyzing these results, we can conclude that the average improvements in the PSNR values are not significant in the majority of the sequences. An exception is the Carphone sequence that contains many clear-defined edges and, hence, the inclusion of an edge-dependent algorithm improves its performance in more than 0.2 dB.

Although the improvements in numerical results are not significant, the visual inspection of the images discovers relevant advantages of 'proposal Chapter 4', as can be corroborated in Figures 4.34, 4.35, and 4.36. These images correspond to video sequences and the advantages in the reconstruction of edges are especially evident in the areas depicted by ellipses for the Carphone and Salesman sequences (see Figure 4.34) and the Phantom Menace sequence (see Figure 4.35). The result on a video sequence called Calen (with a high number of sharp details) is shown in Figure 4.36.

Furthermore, some images that correspond to film sequences are included to illustrate the advantage of the new spatial interpolator (see Figures 4.37, 4.38, and 4.39). They show the superior performance in the reconstruction of moving edges depicted with black ellipses. In Figures 4.37 and 4.38, two zooms of a de-interlaced frame from the Tokyo sequence are shown. Finally, an area of the Bodyguard sequence with a moving clear edge is shown in Figure 4.39.

4.3 Conclusions

Since the human visual system is especially sensitive to the edges of an image, several edge-dependent spatial interpolators have been proposed in the literature so as to de-interlace them correctly, avoiding the staircase effect of *line average*. Edges, as well as motion, are usually fuzzy and, hence, crisp algorithms such as ELA (Edge-dependent Line Average) introduce mistakes in unclear and/or ambiguous edges. The first fuzzy ELA proposed in this chapter (named 'Basic Fuzzy ELA') represents a trade-off between *line average* and ELA: when edges are clearly detected, interpolation is performed as well as ELA (improving *line average* results), while when edges are not clear, interpolation mixes ELA with *line average* or even only applies *line average* (avoiding mistakes of ELA).

To further improve 'Basic Fuzzy ELA', three variations have been analyzed: one focused on ensuring the presence of an edge in areas with non-clearly defined edges (named 'Recursive Fuzzy ELA') and two others focused on increasing (from three to five) the number of directions where looking for edges (named 'Fuzzy-ELA 5+5' and 'Improved Fuzzy-ELA 5+5'). After extensive analysis, the conclusion is that 'Improved Fuzzy-ELA 5+5' beats the other edge-dependent interpolators (in terms of quantitative PSNR values and qualitative visual sensation) in both clear and unclear edges.

Using the proposed fuzzy spatial interpolator in the fuzzy motion-adaptive algorithm described till now clearly improves the de-interlaced results in those areas of the image with large motion. However, small improvements are obtained in zones where the level of motion is small because a poor temporal interpolator is being used there. Hence, the next step given in the following chapter is to improve the performance of the temporal interpolator.

References

1. Doyle, T., Looymans, M.: Progressive scan conversion using edge information. In: Chiariglione, L. (ed.) Signal Processing of HDTV, II., pp. 711–721. Elsevier Science Publishers, Amsterdam (1990)
2. Lee, M.H., Kim, J.H., Lee, J.S., Ryu, K.K., Song, D.: A new algorithm for interlaced to progressive scan conversion based on directional correlations and its IC design. IEEE Trans. on Consumer Electronics 40(2), 119–129 (1994)

3 Kuo, C.J., Liao, C., Lin, C.C.: Adaptive interpolation technique for scanning rate conversion. IEEE Trans. on Circuits and Systems for Video Technology 6(3), 317–321 (1996)
4 Lee, H.Y., Park, J.W., Bae, T.M., Choi, S.U., Ha, Y.H.: Adaptive scan rate up-conversion system based on human visual characteristics. IEEE Trans. on Consumer Electronics 46(4), 999–1006 (2000)
5 Salonen, J., Kalli, S.: Edge adaptive interpolation for scanning rate conversion. In: Signal Processing of HDTV IV, pp. 757–764. Elsevier, Amsterdam (1993)
6 Simonetti, R., Filisan, A.P., Carrato, S., Ramponi, G., Sicuranza, G.: A deinterlacer for IQTV receivers and multimedia applications. IEEE Trans. on Consumer Electronics 39(3), 234–240 (1993)
7 De Haan, G., Lodder, R.: De-interlacing of video data using motion vector and edge information. In: Proc. IEEE Int. Conf. on Consumer Electronics (ICCE), Los Angeles, USA (June 2002)
8 Chang, Y.L., Lin, S.F., Chen, L.G.: Extended intelligent edge-based line average with its implementation and test method. In: Proc. IEEE Int. Symposium on Circuits and Systems (ISCAS), Vancouver (Canada), May 2004, vol. 2, pp. 341–344 (2004)
9 Yoo, H., Jeong, J.: Direction-oriented interpolation and its application to de-interlacing. IEEE Trans. on Consumer Electronics 48(4), 954–962 (2002)

Chapter 5
Fuzzy Motion-Adaptive De-Interlacing with Smart Temporal Interpolation

Abstract. The performance of a motion-adaptive de-interlacing algorithm is directly dependent on the quality of the spatial and temporal interpolators. As shown in Chapter 4, the inclusion of an edge-dependent algorithm clearly improves the results in moving areas of the image, whereas small improvements are obtained in zones where the level of motion is inferior since a combination of the spatial and temporal interpolators is carried out in these cases. Similarly, the performance of the motion-adaptive de-interlacing algorithm relies partially on the quality of the temporal interpolator in pixels with an intermediate level of motion and the dependence is complete in static areas. The aim of this Chapter is to improve the temporal interpolator used in the fuzzy motion-adaptive de-interlacing algorithm developed in the previous Chapters of this book. The temporal technique used in the de-interlacing process could be selected according to the origin of image sequences. In this sense, pull-down material is increasingly combined with interlaced video such as computer generated moving and/or animated imagery or scrolling texts. Unfortunately, knowledge of the picture repetition pattern is not included in the transmission, and this information is highly relevant for de-interlacing. For instance, de-interlacing is perfectly performed by weaving the correct fields of pull-down material. This algorithm is the best due to its simplicity, however, if the correct field is not selected, an annoying artifact known as 'feathering' appears (see Figure 5.1). Different detectors have been proposed in the literature to identify the field-pairs originated by the pull-down process from the same film image. The primitive approaches perform a global identification for the entire field, that is, a control signal is activated to denote the presence of film material. One of these detectors, called zero-vector matching detectors, try to match the zero motion vector on a previous field. This kind of detectors normally use two kinds of signals: a first one to calculate the frame similarity and a second one to measure the field similarity. Based on the analysis of both similarity metrics, control signals are generated, which indicate the mode of the video signal, i.e. video or film, and the type and phase of the film mode to determinate the image's position in the 3:2 or 2:2 pull-down pattern. This approach is described in [1], and it is used by the majority of current film detectors. However, there are some drawbacks in the performance of these conventional film detectors since the

P. Brox et al.: Fuzzy Logic-Based Algorithms for Video De-Interlacing, STUDFUZZ 246, pp. 143–168.
springerlink.com

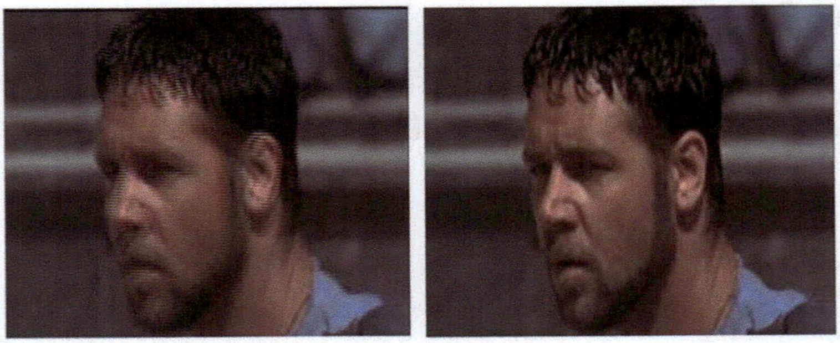

Fig. 5.1. Feathering effect (left) that appears when picture-repetition is assumed incorrectly.

sum of absolute differences and its comparison with an unique threshold value is not suitable. Firstly, an unique sum does not properly measure the level of motion. For example, in the case that the majority of parts in the image are static but there are moving parts of small dimensions, a low value of the global sum is achieved. This ignores the moving areas and, hence, feathering effect appears in this moving area. A second disadvantage is the determination of the threshold value. Its selection is extremely difficult since the properties of the image are very variable. Other approaches try to analyze the frame characteristics to find out the repetition pattern [2]-[5]. Among them, several proposals have been reported in the literature to identify jagged edges in frames due to 'feathering' effect [2],[3]. Other detectors study the location of edges in the frames [4]. The idea is that, if two frames are similar, edges should be at the same spatial position. Therefore, the analysis of edges position can reveal the picture repetition pattern. Finally, a motion vector based approach has been proposed in [5]. The sum of the length of the motion vectors should indicate if two fields are identical or not, since fields from the same frame should provide a similar value of the sum. In recent literature, the research works in the area of film-detection can be categorized into two groups. A first one is focused on the increase of the robustness in the detection of pattern repetition. This is especially relevant since wrong mode decisions can produce annoying artifacts. In [6], the number of wrong film mode decisions are reduced by using a new motion detection scheme called the 'Arrow' detector. It reduces the problem of interpreting vertically detailed areas as motion. The method proposed in [7] also obtains a considerable robustness improvement by using a layered structure that separates the film-mode source from interlaced TV signal to guarantee the correction of weaving sequence. The second group of research works is focused on the development of film-mode detectors that work locally. This kind of algorithms is very demanded due to the high increase of TV material that combines images from different origins in a single field, which is known as hybrid material. None of the techniques previously cited can locally deal with hybrid material, because all of them detect a single mode for the entire field. The proposal described in [8] can assign different modes in a single field. This algorithm firstly implements an image segmentation process and, after that, one mode decision is made for each detected object. In [9], an adaptive interpolation is performed by weighting between *field insertion* and other interpolation algorithm for de-interlacing purpose. The weighting factors are the output of a fuzzy system, and they are obtained by analyzing as inputs of the system the intra and inter-field signal differences concerning the current pixel along a set of pre-determinated directions. Based on the idea of film-mode detectors, an effective temporal interpolator is presented in this Chapter. The aim is to detect the most adequate combination of neighboring pixels in the temporal domain, so that, weaving is applied with the correct field when picture repetition appears. Since this strategy is locally performed, the algorithm can deal with any kind of sequences: film, hybrid, and video sequences. This Chapter is organized as follows: Section 5.1 describes the new approach, which is able to adapt the strategy of interpolation to the presence of repeated areas in consecutive fields. The advantages of including the new temporal interpolator are shown in Section 5.2 by de-interlacing video, hybrid, and film sequences. These results are compared with those obtained with the previous versions of the algorithm in Section 5.3, whereas Section 5.4 summarizes a comparison with MC de-interlacing algorithms. Finally, some conclusions are expounded in Section 5.6.

5.1 A Smart Temporal Interpolator

In order to detect repeated areas in the fields, a measure of dissimilarity between two consecutive fields is performed. Taking advantage of previous experience, dissimilarity between two consecutive fields is calculated by using a bi-dimensional convolution that requires the previous field to be de-interlaced. Hence, the first field in the sequence should be de-interlaced by a spatial interpolator method (in our case, the 'Improved Fuzzy-ELA 5+5' algorithm is used). Since the previous de-interlaced field is now available, a new strategy can be used to measure motion by evaluating three instead of four consecutive fields. Therefore, the following bi-dimensional convolution of field differences is used:

$$motion(x,y,t) = \frac{\Sigma\Sigma M_{(i,j)}C_{(i,j)}}{\Sigma\Sigma C_{(i,j)}} \tag{5.1}$$

where $M_{(i,j)}$ are the elements of a matrix, M, of difference values among three consecutive fields and $C_{(i,j)}$ are the coefficients of the convolution mask. For instance, the matrices $M_{(i,j)}$ and $C_{(i,j)}$ when the index 'i' equals to 3 and 'j' equals to 5 are as follows:

$$C = \begin{pmatrix} 1\,2\,3\,2\,1 \\ 1\,3\,4\,3\,1 \\ 1\,2\,3\,2\,1 \end{pmatrix} \tag{5.2}$$

$$M = \begin{pmatrix} M_{(-2,-1)} & M_{(-1,-1)} & M_{(0,-1)} & M_{(1,-1)} & M_{(2,-1)} \\ M_{(-2,0)} & M_{(-1,0)} & M_{(0,0)} & M_{(1,0)} & M_{(2,0)} \\ M_{(-2,1)} & M_{(-1,1)} & M_{(0,1)} & M_{(1,1)} & M_{(2,1)} \end{pmatrix} \tag{5.3}$$

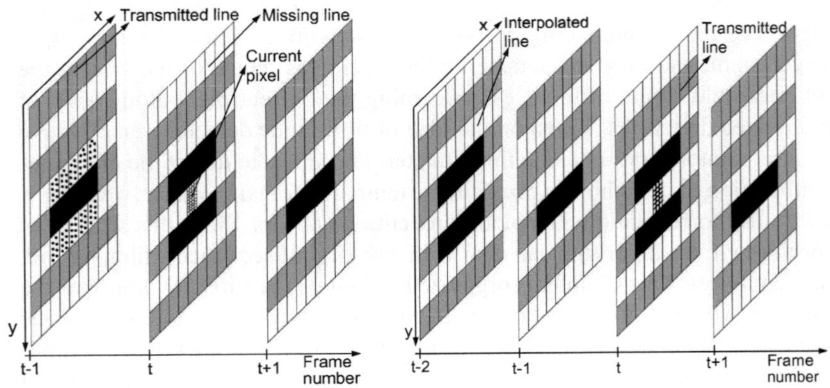

(a) The black and spotted pixels are used in the calculus of the new motion measure

(b) The black pixels are used in the algorithm proposed in Chapter 3

Fig. 5.2. Pixel involved in the calculation of the bi-dimensional convolution.

The elements of the matrix $M_{(i,j)}$ are calculated as follows:

$$M_{(i,0)} = M(x+i,y,t) = \frac{|I(x+i,y,t+1) - I(x+i,y,t-1)|}{2} \tag{5.4}$$

$$M_{(i,-1)} = M(x+i,y-1,t) = \frac{|I(x+i,y-1,t) - I_p(x+i,y-1,t-1)|}{2} \tag{5.5}$$

$$M_{(i,1)} = M(x+i,y+1,t) = \frac{|I(x+i,y+1,t) - I_p(x+i,y+1,t-1)|}{2} \tag{5.6}$$

where I_p are the interpolated luminances of the pixels in the previous field.

Figure 5.2(a) shows how the new measure of *motion* involves the pixels from three consecutive fields. This new measure of motion is the input of the motion-adaptive fuzzy system described in Chapter 3, which combines a temporal with a spatial interpolator. The detailed pixels in Figure 5.2(a) are used when '$i = 3$' and '$j = 5$'. The measure of *motion* in Chapter 3 evaluates a temporal aperture of four instead of three fields as shown in Figure 5.2(b).

Since detection of repeated areas in the fields is relevant for a smart temporal interpolation, the following dissimilarity measure is proposed as the input of a new temporal interpolator:

$$dissimilarity(x,y,t) = \frac{\Sigma\Sigma M_{(i,j)} C'_{(i,j)}}{\Sigma\Sigma C'_{(i,j)}} \tag{5.7}$$

which, similarly to motion measure in (5.1), is a bi-dimensional convolution but only involving pixels in the current and the previously deinterlaced fields, because the matrix of weights, C', has null its even rows, for instance, C' can be as follows:

$$C' = \begin{pmatrix} 1\,2\,3\,2\,1 \\ 0\,0\,0\,0\,0 \\ 1\,2\,3\,2\,1 \end{pmatrix} \tag{5.8}$$

The influence of *dissimilarity* in selecting the kind of temporal interpolation is evaluated by considering the following fuzzy rules:

1. If *dissimilarity* between the fields *(t-1)* and *(t)* is *SMALL* (S), the most adequate interpolated value is obtained by selecting the pixel value in the previous field at the same spatial position ($I(x,y,t-1)$).
2. On the contrary, if *dissimilarity* is *LARGE* (L), the pixel value in the previous field is not a good choice and is better to bet on the pixel in the next field ($I(x,y,t+1)$).

Figure 5.3(a) summarizes the rule base of this new fuzzy system. The shape of membership functions to model the fuzzy concepts *SMALL* and *LARGE* are shown in Figure 5.3(b). The output of this fuzzy system is given by the following expression:

$$I_T = \gamma_1 \cdot I(x,y,t-1) + \gamma_2 \cdot I(x,y,t+1) \tag{5.9}$$

where γ_i is the activation of each rule in the rule base of Figure 5.3(a).

Rule	Antecedent	Consequent
1.	*dissimilarity (x,y,t)* is *SMALL*	*I(x,y,t-1)*
2.	*dissimilarity (x,y,t)* is *LARGE*	*I(x,y,t+1)*

(a) Rule base for the temporal interpolator

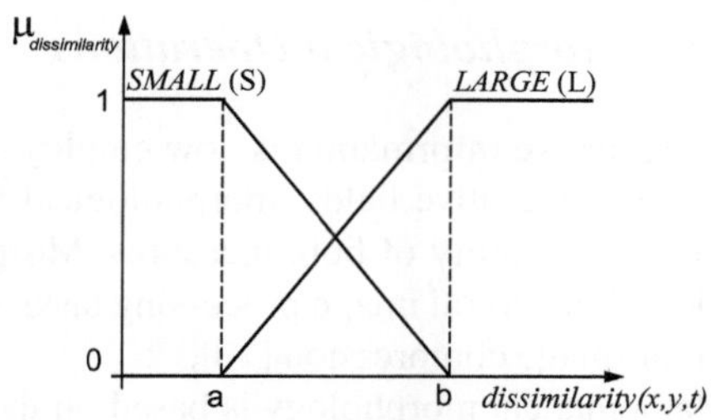

(b) Membership functions

Fig. 5.3. Fuzzy system for the temporal interpolator.

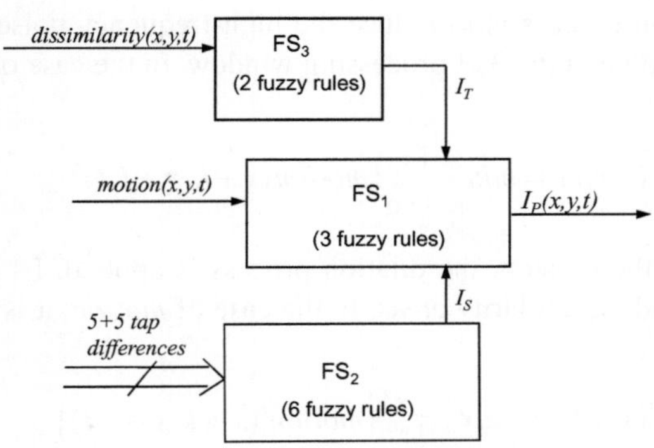

Fig. 5.4. Block diagram of the global algorithm.

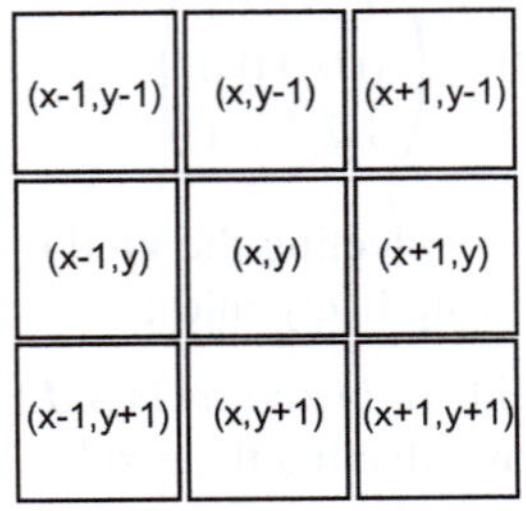

Fig. 5.5. 3x3 processing window.

The block diagram of the resulting motion-adaptive de-interlacing algorithm is shown in Figure 5.4. 'FS_1' symbolizes the fuzzy system to implement the motion-adaptive de-interlacing presented in Chapter 3. The block 'FS_2' represents the fuzzy system, presented in Chapter 4, to implement the 'Improved 5+5 Fuzzy-ELA' algorithm. The third fuzzy system, introduced in this Chapter, performs the temporal interpolation and is denoted as 'FS_3'.

5.1.1 Morphological Operations

Since recursive information is now employed to measure *motion* and *dissimilarity* between consecutive fields, morphological operators have been introduced to improve the reliability of both measures. Morphological operators have been widely employed in several image processing tasks such as noise removal, image enhancement, or image compression [10].

Mathematical morphology is based on the idea that certain features of an image (in our case, motion and dissimilarity) are a property of an area of the image rather than a property of each pixel. Typically, two basic operators are applied: erosion and dilation, which usually consider a 3x3 square processing window around the current pixel, as shown in Figure 5.5.

The aim of erosion process is to reduce the high-frequency noise by selecting the minimum pixel value in the 3x3 processing window. In the case of *motion*, it is calculated as follows:

$$motion'(x,y,t) = min_{\substack{k=\pm1,0 \\ l=\pm1,0}} \{motion(x+k,y+l,t)\} \tag{5.10}$$

After performing the erosion, the dilation process is applied. Its purpose is to extend the motion and dissimilarity areas. In the case of *motion*, it is calculated as follows:

$$motion''(x,y,t) = max_{\substack{k=\pm1,0 \\ l=\pm1,0}} \{motion'(x+k,y+l,t)\} \tag{5.11}$$

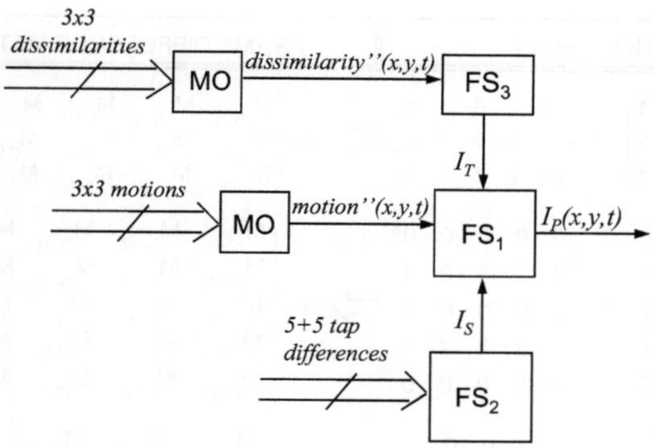

Fig. 5.6. Block diagram of the global algorithm including morphological operations, MO.

With the inclusion of these morphological operations, the final block diagram of our proposed de-interlacing algorithm is shown in Figure 5.6.

5.2 Performance of the Proposed Algorithm

Similarly to the study described in Chapter 3[1], several options for the matrix M and its corresponding matrices of weights C and C' have been analyzed (see Figure 5.7). As can be seen, the simplest approach uses a matrix C5 of size 3x1 whereas the most complex one, C1, has a size of 5x5. The criterion followed to select the values of C is to assign the highest weight to the position placed on the current pixel $(i, j = 0)$. The values of the rest of weights are reduced according to a radial symmetry centered on the current pixel as detailed in Chapter 3. The matrices C' are obtained from the matrices C by setting null values to the even rows. The performance of all the alternatives presented in Figure 5.7 are studied in this Section.

The six options considered in Figure 5.7 have been used to de-interlace the battery of video and film sequences. As shown in Table 5.1, the approach with the convolution masks C4 and C4' obtains the highest PSNR results in both types of material. Among the rest of proposals, the simplest option (C5, C5') achieves the second highest PSNR results very frequently. The results in Table 5.1 also show that the proposal with the most complex convolution masks (C1 and C1') presents an irregular behavior. It is always inferior to the algorithms with the convolution masks (C4, C4') and (C5, C5'), and often better than other options.

Since the convolution masks {C4, C4'} obtain the best results, they have been selected to study how the algorithm performs with the use of morphological operators. The latest column in Table 5.1 shows how the results of the algorithm with the convolution masks {C4, C4'} are further improved by including morphological operations (MO).

[1] In fact, the matrices of weights C are similar to the ones used in Chapter 3 except for three values of the matrix C. They have been changed in order to ease the normalization of the matrix C since the sum of the elements is now a power of two.

CONVOLUTION MASKS		FRAME DIFFERENCE MATRICES
$C=\begin{pmatrix}1&2&3&2&1\\2&3&4&3&2\\1&2&3&2&1\end{pmatrix}$	$C'=\begin{pmatrix}1&2&3&2&1\\0&0&0&0&0\\1&2&3&2&1\end{pmatrix}$	$\Rightarrow M=\begin{pmatrix}M_{(-2,-1)}&M_{(-1,-1)}&M_{(0,-1)}&M_{(1,-1)}&M_{(2,-1)}\\M_{(-2,0)}&M_{(-1,0)}&M_{(0,0)}&M_{(1,0)}&M_{(2,0)}\\M_{(-2,1)}&M_{(-1,1)}&M_{(0,1)}&M_{(1,1)}&M_{(2,1)}\end{pmatrix}$
$C1=\begin{pmatrix}1&1&2&1&1\\1&2&3&2&1\\2&3&4&3&2\\1&2&3&2&1\\1&1&2&1&1\end{pmatrix}$	$C1'=\begin{pmatrix}0&0&0&0&0\\1&2&3&2&1\\0&0&0&0&0\\1&2&3&2&1\\0&0&0&0&0\end{pmatrix}$	$\Rightarrow M=\begin{pmatrix}M_{(-2,-2)}&M_{(-1,-2)}&M_{(0,-2)}&M_{(1,-2)}&M_{(2,-2)}\\M_{(-2,-1)}&M_{(-1,-1)}&M_{(0,-1)}&M_{(1,-1)}&M_{(2,-1)}\\M_{(-2,0)}&M_{(-1,0)}&M_{(0,0)}&M_{(1,0)}&M_{(2,0)}\\M_{(-2,1)}&M_{(-1,1)}&M_{(0,1)}&M_{(1,1)}&M_{(2,1)}\\M_{(-2,2)}&M_{(-1,2)}&M_{(0,2)}&M_{(1,2)}&M_{(2,2)}\end{pmatrix}$
$C2=\begin{pmatrix}1&2&1\\2&3&2\\3&4&3\\2&3&2\\1&2&1\end{pmatrix}$	$C2'=\begin{pmatrix}0&0&0\\2&3&2\\0&0&0\\2&3&2\\0&0&0\end{pmatrix}$	$\Rightarrow M=\begin{pmatrix}M_{(-1,-2)}&M_{(0,-2)}&M_{(1,-2)}\\M_{(-1,-1)}&M_{(0,-1)}&M_{(1,-1)}\\M_{(-1,0)}&M_{(0,0)}&M_{(1,0)}\\M_{(-1,1)}&M_{(0,1)}&M_{(1,1)}\\M_{(-1,2)}&M_{(0,2)}&M_{(1,2)}\end{pmatrix}$
$C3=\begin{pmatrix}1&2&1\\2&4&2\\1&2&1\end{pmatrix}$	$C3'=\begin{pmatrix}1&2&1\\0&0&0\\1&2&1\end{pmatrix}$	$\Rightarrow M=\begin{pmatrix}M_{(-1,1)}&M_{(0,-1)}&M_{(1,-1)}\\M_{(-1,0)}&M_{(0,0)}&M_{(1,0)}\\M_{(-1,1)}&M_{(0,1)}&M_{(1,1)}\end{pmatrix}$
$C4=\begin{pmatrix}0&1&0\\1&2&1\\0&1&0\end{pmatrix}$	$C4'=\begin{pmatrix}0&1&0\\0&0&0\\0&1&0\end{pmatrix}$	$\Rightarrow M=\begin{pmatrix}M_{(-1,1)}&M_{(0,-1)}&M_{(1,-1)}\\M_{(-1,0)}&M_{(0,0)}&M_{(1,0)}\\M_{(-1,1)}&M_{(0,1)}&M_{(1,1)}\end{pmatrix}$
$C5=\begin{pmatrix}1\\2\\1\end{pmatrix}$	$C5'=\begin{pmatrix}1\\0\\1\end{pmatrix}$	$\Rightarrow M=\begin{pmatrix}M_{(0,-1)}\\M_{(0,0)}\\M_{(0,1)}\end{pmatrix}$

Fig. 5.7. Several options of matrices M and its corresponding matrices of weights C and C'.

The advantages introduced by the inclusion of the morphological operations can be seen with the snapshots of the hybrid TMF sequence[2] (see Figure 5.8). This sequence is an interlaced video clip (2:2 pull-down mode) with an overlay containing a ticker-tape video text. It also contains stationary areas (around the clock and the TMF-logo).

Film area in the snapshot of the left part in Figure 5.8 agrees with the previous image, whereas film area of the snapshot in the right part of Figure 5.8 belongs to a field in which film area is similar to the next field in the sequence.

In Figures 5.9 and 5.10, black color represents the static area whereas white represents the moving areas according to the measure of motion in equation (5.1). In Figures 5.11 and 5.12, black color means that there is similarity between the current and previous field whereas white means a high difference between the current field and the previous one.

[2] TMF is not included in Table 5.1 since its corresponding progressive material is not available. This sequence was provided by 'Philips Research Laboratories' in Eindhoven (The Netherlands).

Table 5.1. PSNR results (in dBs) achieved by the options in Figure 5.7 for video and film sequences

Convolution masks	{C5, C5'}	{C4, C4'}	{C3, C3'}	{C2, C2'}	{C1, C1'}	{C, C'}	{C4, C4'}+MO
News	38.78	39.58	38.41	38.63	38.74	38.12	39.91
Mother	42.11	42.45	41.77	41.89	41.92	41.56	42.37
Carphone	35.09	35.16	35.02	35.01	34.76	34.72	35.69
Missa	40.81	40.86	40.56	40.55	39.57	40.54	40.82
Paris	35.87	36.38	35.49	35.64	35.77	35.19	36.91
Salesman	38.35	38.78	38.49	38.61	38.71	38.41	38.94
Trevor	37.63	37.69	37.53	37.56	37.54	37.43	37.98
Phantom Menace	32.98	33.12	32.92	32.94	32.91	32.78	34.92
DieAnother Day	40.45	41.38	40.44	40.44	41.03	40.11	42.76
Fire Rose	40.97	40.98	40.85	40.81	40.63	40.79	41.45
Fargo1	37.81	37.93	37.75	37.46	37.41	37.52	38.96
Fargo2	41.01	41.86	40.99	41.31	41.49	40.71	43.65
Tokyo	36.97	37.04	36.82	36.74	37.04	36.71	38.17
Bodyguard	38.88	39.61	38.87	39.01	39.16	38.77	41.11

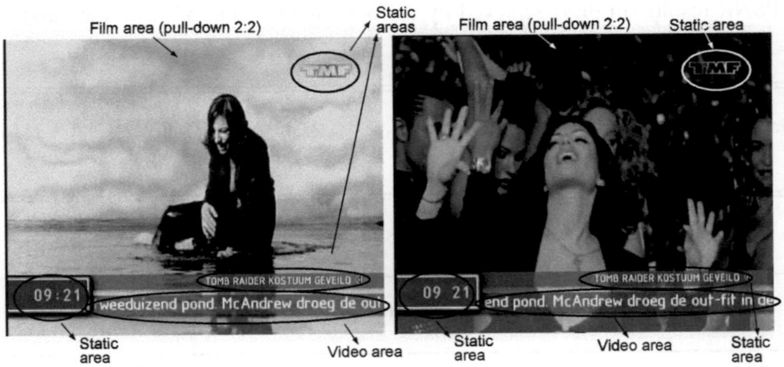

Fig. 5.8. Snapshots of the hybrid TMF sequence.

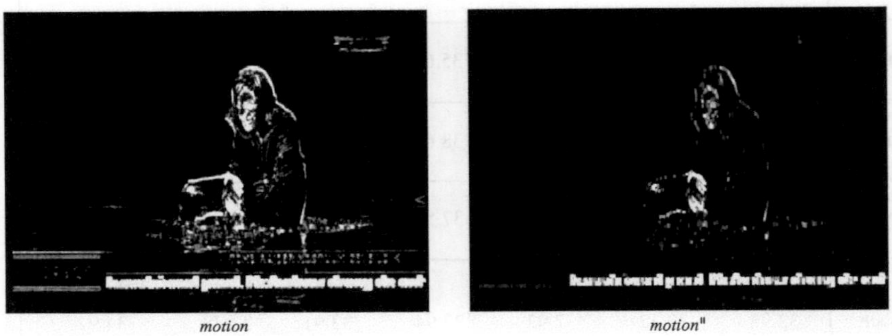

(a) Measure of *motion* without MO

(b) Measure of *motion* with MO

Fig. 5.9. Advantages of using morphological operators in the calculus of *motion* measure for Figure 5.8(left).

As shown in Figures 5.9(b) and 5.10(b), morphological operations considerably eliminate false cases of motion in stationary areas (for instance around TMF logo, clock time, and letters of 'Tomb Raider'). Furthermore, as shown in Figure 5.11(b), they improve the robustness of *dissimilarity* measure since the previous and current film parts in the sequence are similar while only the video text is not similar. The better reliability of *dissimilarity* measure is also corroborated by Figure 5.12(b), in which film area is different from the previous one.

The improvements can also be corroborated by the visual inspection of the de-interlaced images, as shown in the zooms of Figure 5.13. To show up the advantages of the final approach, Figure 5.13 also contains the results achieved by other de-interlacing algorithms with less or similar computational cost.

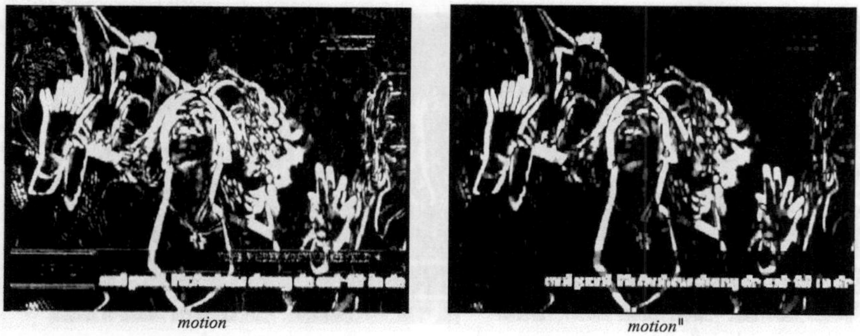

(a) Measure of *motion* without MO (b) Measure of *motion* with MO

Fig. 5.10. Advantages of using morphological operators in the calculus of *motion* measure for Figure 5.8(right).

(a) Measure of *dissimilarity* without MO (b) Measure of *dissimilarity* with MO

Fig. 5.11. Advantages of using morphological operators in the calculus of *dissimilarity* measure for Figure 5.8(left).

5.3 Evolution of the Fuzzy De-Interlacing Proposals

Our final algorithm for video de-interlacing has been obtained after the subsequent improvements presented in Chapter 2, 3, and 4. The evolution of the results is shown in Tables 5.2 and 5.3, in which the 'final proposal' corresponds to the proposal with the convolution masks {C4, C4'} with the inclusion of morphological operations (MO). As can be seen from these results, the final proposal always achieves the highest PSNR results. Its improvements versus the rest of proposals are significant, more than '0.5 *dBs*', in the majority of the sequences.

The performance of these algorithms is detailed field-by-field in three of the sequences: Carphone, Salesman, and Fargo2. The results in Carphone sequence outperform the performance of the rest of de-interlacing methods in the majority of

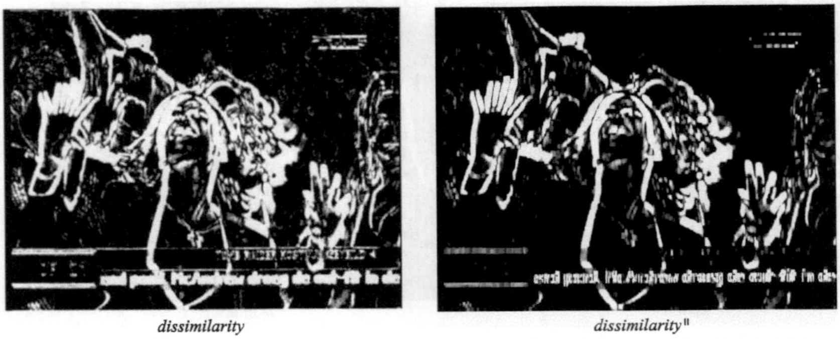

(a) Measure of *dissimilarity* without MO (b) Measure of *dissimilarity* with MO

Fig. 5.12. Advantages of using morphological operators in the calculus of *dissimilarity* measure for Figure 5.8(right).

Fig. 5.13. The definite approach versus other de-interlacing methods.

fields (see Figure 5.14). Particularly, the preliminary algorithm called 'Algorithm in Chapter 2' is significantly improved.

In Salesman sequence, the final approach also achieves the lowest errors but the improvements in error measures are smaller as shown in Figure 5.15. The major advantages are obtained in the Fargo2 sequence as expected. Since this sequence corresponds to film material, the inclusion of the fuzzy system to detect material from picture repetition patterns should be noticed. As shown in Figure 5.16, the 'final proposal' considerably reduces the error values in odd numbers of the sequence, and its performance is barely affected by the parity of the field number.

Table 5.2. PSNR results (in dBs) achieved by the algorithms described in Chapters 2, 3, 4, and 5 (for video sequences)

	News	Mother	Car-phone	Missa	Paris	Sales-man	Trevor	Phantom Menace	Die AnotherDay
	QCIF	QCIF	QCIF	CIF	CIF	CIF	CIF	DVD PAL	DVD NTSC
Algorithm in Chapter 2	37.51	41.87	34.78	40.18	35.28	38.29	36.69	34.57	42.27
Algorithm in Chapter 3	38.28	41.35	35.05	40.51	35.77	38.45	37.29	34.39	42.35
Algorithm in Chapter 4	38.25	41.36	35.27	40.54	35.68	38.49	37.48	34.43	42.38
Final proposal	39.91	42.37	35.69	40.82	36.91	38.94	37.98	34.92	42.76

Table 5.3. PSNR results (in dBs) achieved by the algorithms described in Chapter 2, 3, 4, and 5 (for film sequences)

	Fire Rose	Fargo 1	Fargo2	Tokyo	Bodyguard
	PAL TV	PAL TV	PAL TV	PAL TV	PAL TV
Algorithm in Chapter 2	41.14	37.33	42.54	37.71	40.28
Algorithm in Chapter 3	39.91	36.74	42.36	38.18	40.79
Algorithm in Chapter 4	39.99	36.65	42.29	38.21	40.61
Final proposal	41.45	38.96	43.65	38.17	41.11

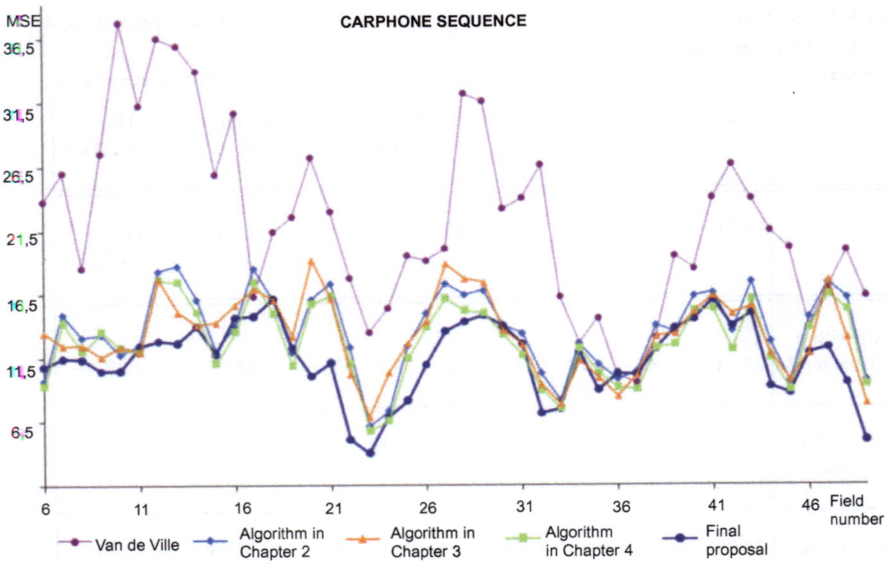

Fig. 5.14. Final approach versus other de-interlacing algorithms in the Carphone sequence.

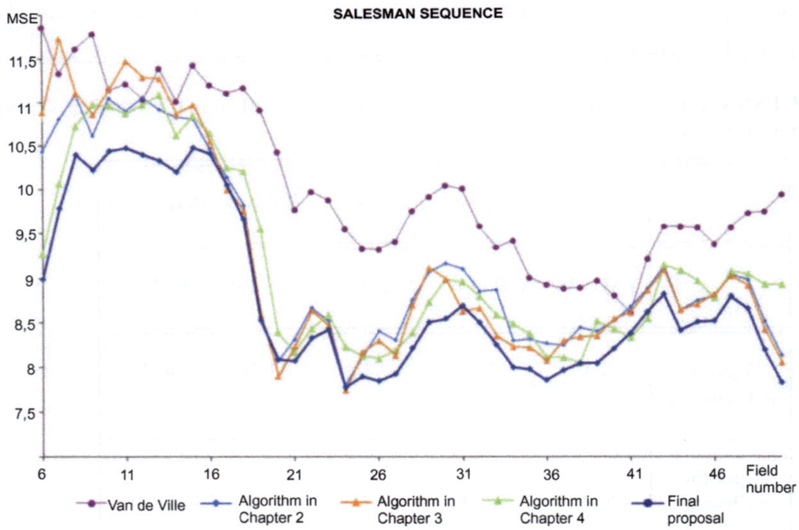

Fig. 5.15. Final approach versus other de-interlacing algorithms in the Salesman sequence.

5.4 Comparison with MC De-Interlacing Methods

The final proposal has demonstrated that it is clearly superior to other de-interlacing algorithms with less or similar computational cost. Consequently, its comparison with other kind of algorithms, the MC de-interlacing algorithms, can be interesting.

Table 5.4. PSNR results (in dBs) for MC methods in video sequences

	News	Mother	Car-phone	Missa	Paris	Sales-man	Trevor	Phantom Menace	Die AnotherDay
	QCIF	QCIF	QCIF	CIF	CIF	CIF	CIF	DVD PAL	DVD NTSC
MC field insertion	36.41	39.72	34.63	40.89	34.48	38.32	37.11	42.96	35.39
MC VT-filtering	36.93	40.93	35.61	41.65	32.46	37.39	38.32	39.61	37.22
MC TBP	36.2	39.49	33.65	40.59	34.29	38.16	37.09	42.84	34.92
MC TR	36.16	39.51	34.61	40.85	34.21	38.01	37.38	42.59	34.74
MC AR	39.69	42.65	36.66	41.41	37.43	38.95	38.68	42.74	37.24
GST	36.07	39.44	34.66	40.15	34.12	37.96	37.12	42.81	35.37
Robust GST	39.48	42.43	36.89	41.98	36.71	38.96	38.44	43.06	37.71
GST-2D	36.09	39.48	34.69	40.23	35.16	37.98	37.16	42.81	35.39
Robust GST-2D	39.21	42.23	36.87	42.01	36.38	38.89	38.81	42.89	37.72
Our proposal	39.91	42.37	35.69	40.82	36.91	38.94	37.98	42.89	34.92

Table 5.4 shows the results provided by nine MC de-interlacing algorithms described in Chapter 1 when de-interlacing was introduced. All the MC methods use the 3-D RS matcher to calculate motion vectors. The parameters used by 3-D RS

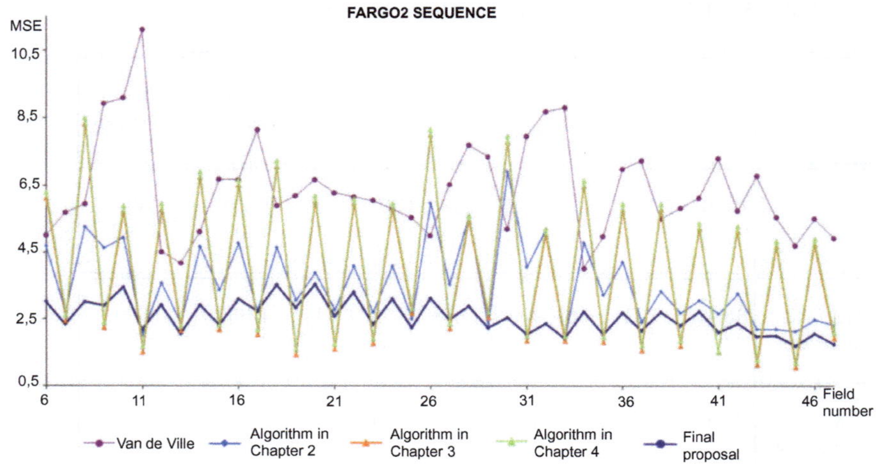

Fig. 5.16. Final approach versus other de-interlacing algorithms in the Fargo2 sequence.

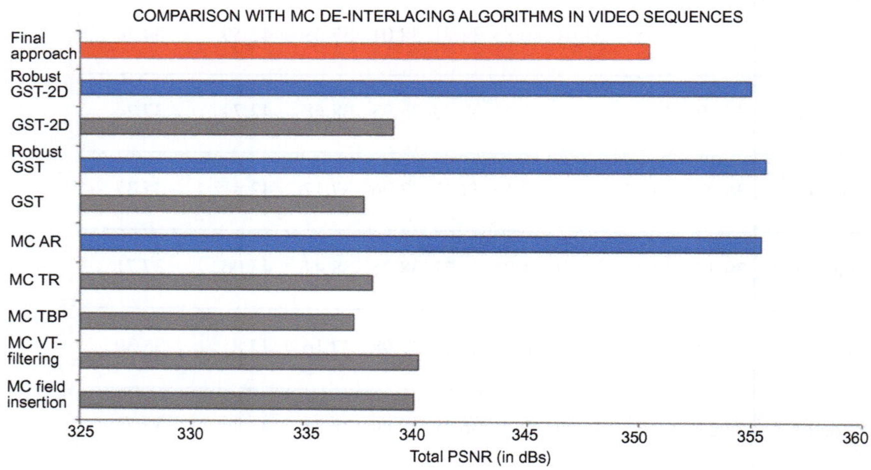

Fig. 5.17. The final approach versus MC de-interlacing algorithms for video sequences.

algorithm are the same in all the MC de-interlacing algorithms. Furthermore, a meandering processing of the pixels and a reverse scan of the fields is performed to improve the quality of this technique [13]. In order to analyze the results better, the sum of these PSNR values for each de-interlacing method is shown in Figure 5.17.

For video sequences, the longest bars correspond to '*MC AR*' and '*Robust GST*' algorithms, which achieve the highest PSNR results in the majority of the sequences, whereas the bar of '*Robust GST-2D*' method is slightly smaller than these two methods. A very interesting result is that our final proposal is clearly superior than a

Table 5.5. PSNR results (in dBs) for MC methods in film sequences

	Fire Rose	Fargo 1	Fargo 2	Tokyo	Bodyguard
	PAL TV	PAL TV	PAL TV	PAL TV	PAL TV
MC field insertion	42.61	42.77	39.44	41.41	35.14
MC VT-filtering	42.36	40.52	39.45	39.91	39.66
MC TBP	42.18	42.16	39.48	41.34	34.89
MC TR	42.82	43.06	38.97	41.33	33.19
MC AR	43.36	41.53	42.85	40.59	41.65
GST	41.85	41.92	39.27	41.55	34.38
Robust GST	41.76	40.07	42.71	39.77	39.94
GST-2D	41.93	41.95	39.28	41.55	34.38
Robust GST-2D	41.75	40.03	42.46	39.58	40.06
Our proposal	41.45	38.96	43.65	38.17	41.11

group of MC methods composed by the following algorithms: *MC field insertion*, *MC VT-filtering*, *MC TBP*, *MC TR*, *GST*, and *GST-2D*.

Table 5.5 shows the average PSNR results obtained after de-interlacing film sequences. It can be seen how our final approach outperforms all the MC methods in Fargo 2 sequence whereas it is quite competitive in Fire Rose and Bodyguard sequences. Figure 5.18 shows the sum of average PSNR achieved by each method in film sequences. The ranking of the methods is the same in video sequences and,

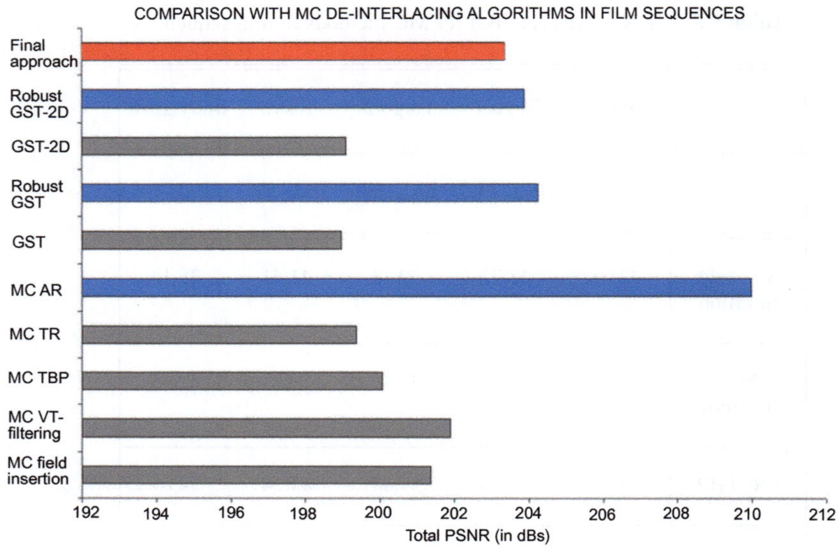

Fig. 5.18. The final approach versus MC de-interlacing algorithms for film sequences.

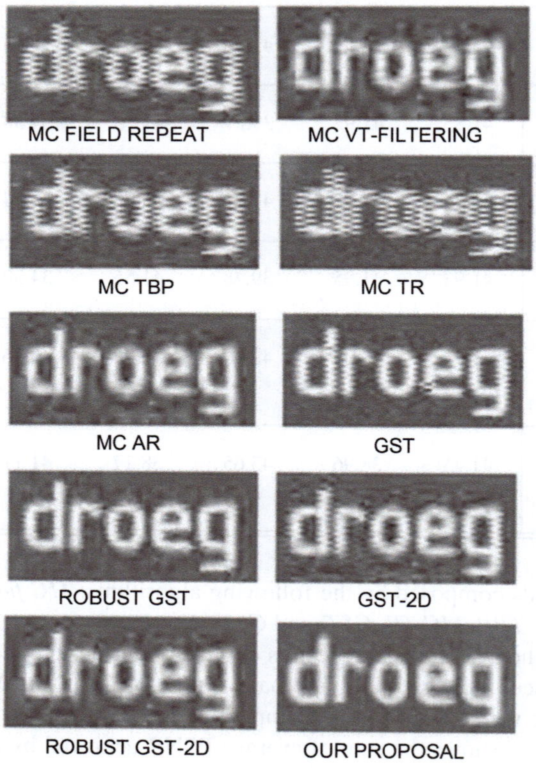

Fig. 5.19. Deinterlaced images obtained with the MC de-interlacing algorithm and our final proposal in the TMF hybrid sequence.

Fig. 5.20. Deinterlaced images obtained with MC de-interlacing algorithms and our final proposal in the Paper 3D hybrid sequence.

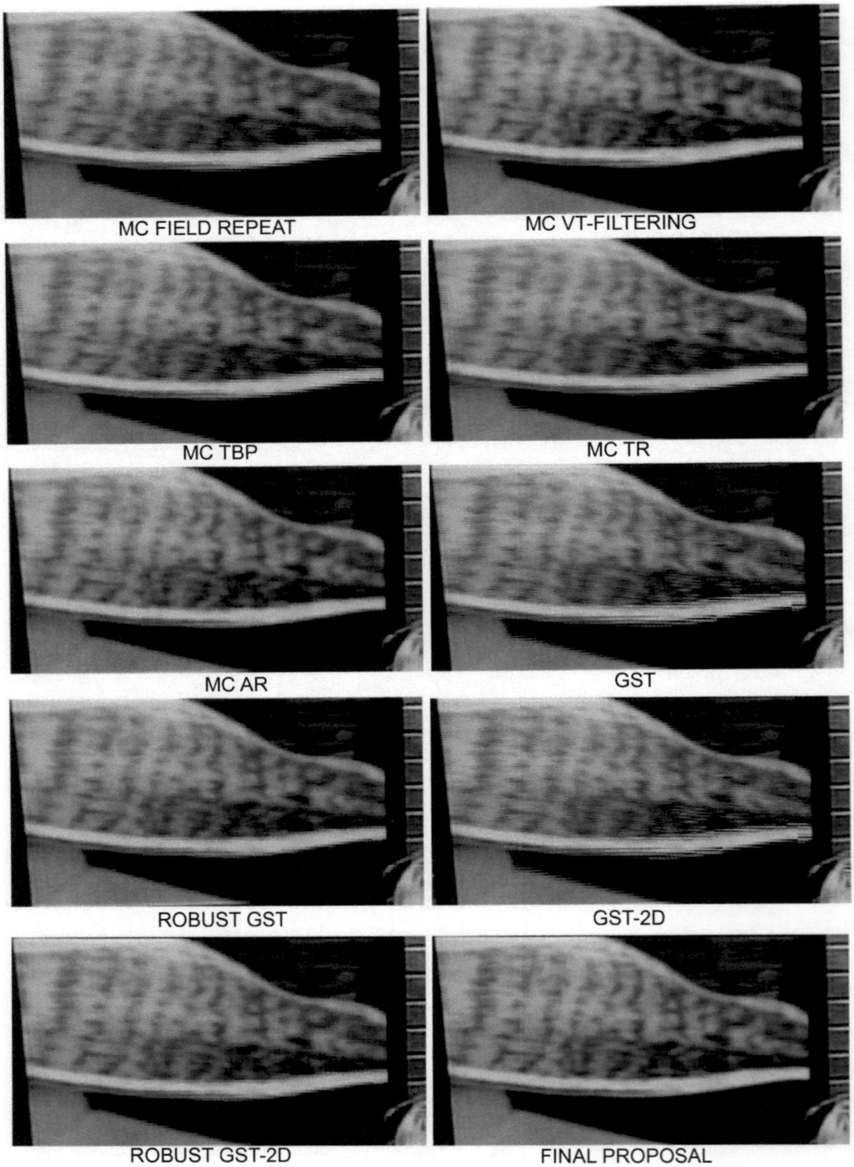

Fig. 5.21. Deinterlaced images obtained with MC de-interlacing algorithms and our final proposal in the Donna hybrid sequence (video area).

again, our final approach achieves the fourth best result. A difference is that in film sequences the group of *MC field insertion*, *MC VT-filtering*, *MC TBP*, *MC TR*, *GST* and *GST-2D* are closer to our approach.

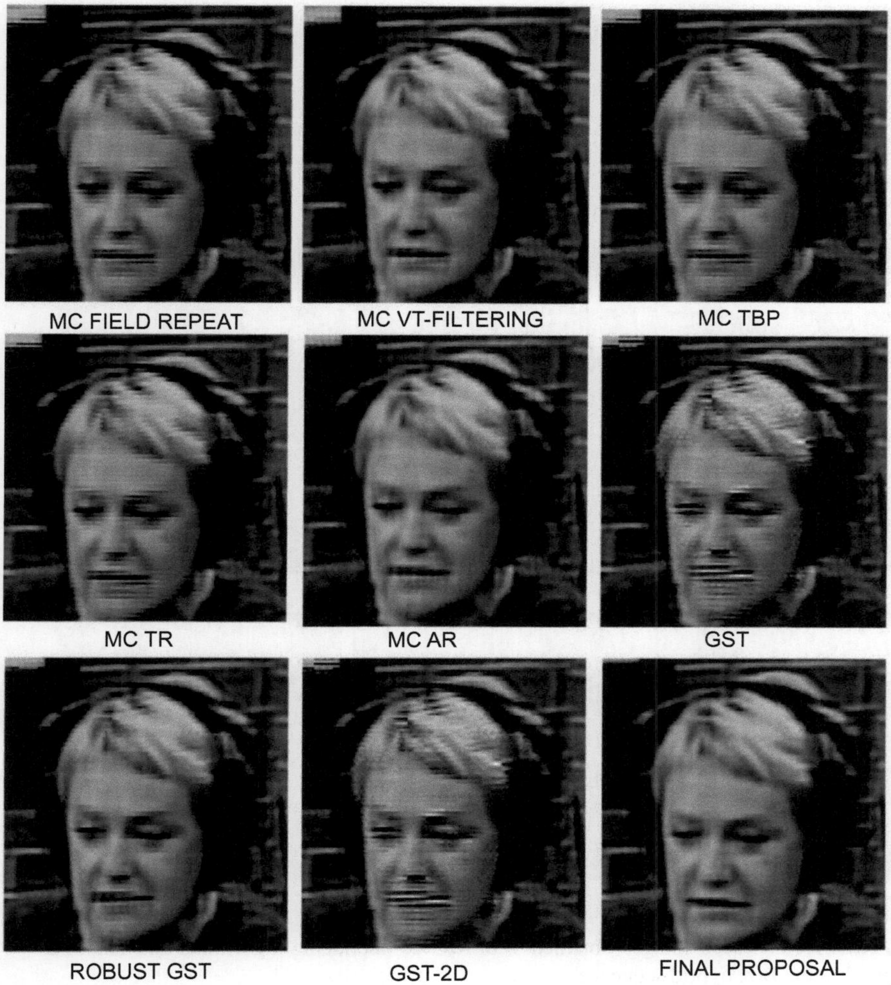

Fig. 5.22. Deinterlaced images obtained with MC de-interlacing algorithms and our final proposal in the Donna hybrid sequence (film area).

The performance of MC de-interlacing algorithms has also been analyzed by de-interlacing four hybrid sequences[3]. Figure 5.19 shows some results after de-interlacing the TMF sequence. Visual inspection of Figure 5.19 corroborates the superior performance of MC AR, Robust GST, and Robust GST-2D over the rest of

[3] These hybrid sequences are not included in Tables 5.4 and 5.5 since their corresponding progressive sequences are not available. These sequences were provided by 'Philips Research Laboratories' in Eindhoven (The Netherlands).

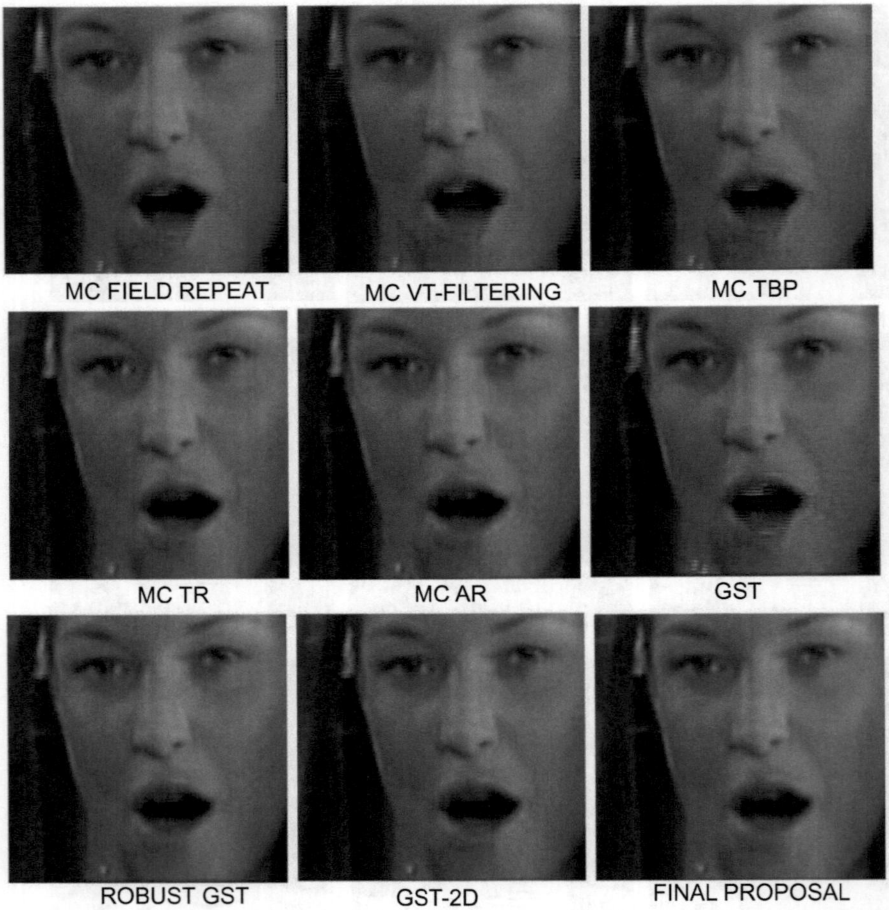

Fig. 5.23. Deinterlaced images obtained with MC de-interlacing algorithms and our final proposal in the Box hybrid sequence.

MC de-interlacing algorithms as well as how our proposal outperforms their results in the reconstruction of the video text.

Figure 5.20 shows a de-interlaced area of the Paper3D hybrid sequence. This sequence contains video material (the letters written on the paper) with a tricky background of film material. Among MC de-interlacing techniques, MC VT-filtering and MC AR achieve the best results. The rest of MC techniques can not appropriately reconstruct the video text and the ghosting effect is clearly appreciated. Our proposal even slightly improves the results of MC VT-filtering and MC AR as shown in Figure 5.20.

Figure 5.21 shows a de-interlaced video area of the Donna hybrid sequence. The scene of the sequence contains a TV in which a video sequence is displayed. Among

Table 5.6. Analysis of storage resources and POs of the final approach

De-interlacing Algorithm	Field memory	Primitive Operations (POs)
MC field insertion	1	529
MC VT filtering	1	536
MC TBP	2	535
MC TR	2	529
MC AR	2	544
GST	1	543
Robust GST	1	555
GST-2D	2	559
Robust GST-2D	2	571
Our approach	3	153

MC de-interlacing techniques, MC AR offers the most attractive results. Again, the results of MC AR is slightly improved by our proposal.

A de-interlaced film area of the Donna sequence is shown in Figure 5.22. This figure corroborates a superior performance of our proposal over MC de-interlacing algorithms in film area due to the inclusion of the new fuzzy system that enables the detection of picture repetition patterns.

Finally, a hybrid sequence called ‘The Box’ has been analyzed. The results are shown in Figure 5.23 for an area of one snapshot that contains film material (the girl’s face). Again two of the MC algorithms (MC VT-filtering and MC AR) obtain

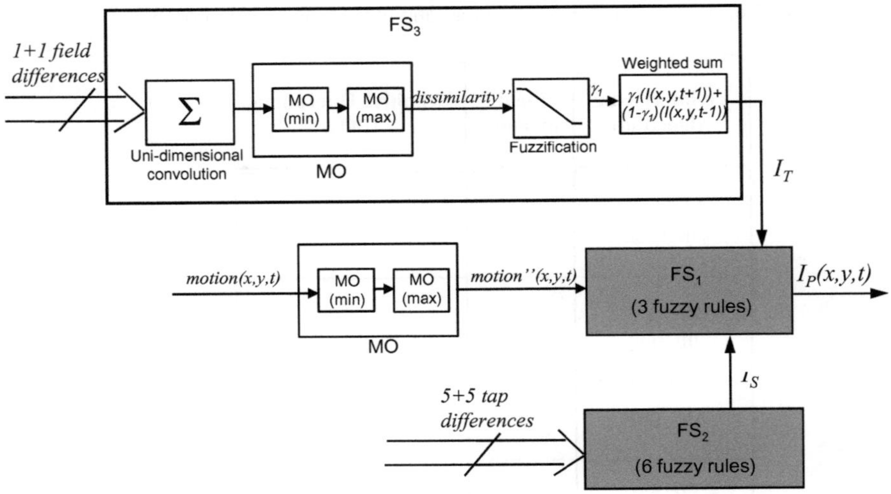

Fig. 5.24. Block diagram of the de-interlacing algorithm.

the best results. This superior performance is specially appreciated around the girl's mouth. Once more time, the results obtained by our proposal are clearly competitive.

Table 5.6 shows the estimation of resources to implement our final approach. Although this method reduces the temporal aperture from four to three fields, the number of field memories required is three since it is necessary to store the values of the interpolated pixels in the previous field (see Figure 5.3(b)).

The number of primitive operations (POs) to perform the final approach is also included in Table 5.6[4]. This number is quite low in comparison with the number of operations required by MC de-interlacing algorithms.

5.5 Conclusions

The simplest temporal interpolator used till this chapter, *field insertion*, introduces annoying artifacts such as feathering when the previous pixel is not a good option. This is especially noticeable in film sequences when the previous field does not correspond to the same but to a different frame, which has motivated an active research field on film-mode detectors. Since our interest is focused on de-interlacing any kind of material: video, film and hybrid (currently very popular due to the explosion of multimedia market), the idea developed is to adapt locally temporal interpolator to the presence of repeated areas in the fields as spatial interpolator was adapted locally to the presence of edges. In this sense, area repetition, as edges and motion, is considered, in general, a local (as happens to video and hybrid sequences) and fuzzy

[4] The calculation of the POs implicitly assumes specific design options to perform the implementations of the algorithms.

feature of the image and, hence, two simple fuzzy rules are proposed to implement a pixel-by-pixel fuzzy selection between previous and posterior pixels. The input of these rules is a measure of dissimilarity between consecutive fields.

Taking advantage of previous experience, dissimilarity between two consecutive fields is performed by a bi-dimensional convolution that requires the previous field to be de-interlaced. Hence, the first field should be de-interlaced by a spatial interpolator (the 'Improved Fuzzy-ELA 5+5'). Once the previous de-interlaced frame is available, another novelty explored in this chapter has been to measure motion by evaluating three instead of four consecutive fields. Since recursive information is now employed, morphological operators (erosion and dilation) have been introduced to improve the reliability of dissimilarity and motion measures.

The resulting algorithm, which exploits fuzzy logic to deal with the presence of motion, edges and repetition, now offers better results (in terms of PSNR and visual inspection) in areas of the images with small and large motion, with clear and unclear edges, and with film and video material mixed.

A very interesting result is that our final proposal is competitive even with motion-compensated algorithms. After an extensive comparative analysis with nine conventional MC algorithms, the results show that three of them (MC AR, Robust GST-1D, and Robust GST-2D) are superior to our final approach in video and film material while the other six methods are far to overcome our results. Clearly, our approach is less costly than MC methods since only the calculation of motion vectors requires more than three times the number of primitive operations required by our proposal.

References

1. Lyon, T.C., Champbell, J.J.: Motion sequence pattern detector for video. Asignee: Faroudja, Yves C., Los Altos Hills, CA. US. United States Patent Office US 4,982,290, January 1 (1991)
2. Correa, C., Schweer, R.: Film mode detection procedure and device. Asignee: Deutsche Thomson-Brandt GMBH, Villingen-Schwennigen (DE). European Patent Office, European Patent EP 0567072B1, July 1 (1998)
3. Lucas, H.Y.W.: Progressive/interlace and redundant field detection for encoder. Applicant: STMICRO Electronics Asia Pacific PTE LTD, Singapore. World Intellectual Property Organization, International Publication Number: WO 00/33579, June 8 (2000)
4. Swan, P.: System and method for reconstructing noninterlaced captured content for display on a progressive screen. Assignee: ATE Technologies, Inc. Thornhill, Canada. United States Patent Ofiice US 6,055,018, April 25 (2000)
5. de Haan, G., Huijgen, H., Biezen, P., Ojo, O.: Method and apparatus for discriminating between movie film and non-movie film and generating a picture signal processing mode control signal. Asignee: U.S. Philips Corporation, New York, USA. United States Patent Office US 5,365,280, November 15 (1994)
6. Dommisse, A.: Film detection for advanced scan rate converters. M. Sc. Thesis, Technische Universiteit Eindhoven (TUE), Eindhoven, The Netherlands (August 2002)
7. Ku, C.-C., Liang, R.-K.: Robust layared film-mode 3:2 pulldown detection/correction. IEEE Trans. on Consumer Electronics 15(4), 1190–1193 (2004)

8. de Haan, G., Wittebrood, R.B.: Recognizing film and video object occuring in parallel in sigle television signal fields. Asignee: Koninklijke Philips Electronics N. V., Eindhoven, NL. United States Patent Office US 6,937,655 (August 2005)
9. He, L., Zhang, H.: Motion object video on film detection and adaptive de-interlace method based on fuzzy logic. Assignee: nDSP Corporation, Campbell, CA, September 258. United States Patent Office US 6,799,168 (2004)
10. Salembier, P., Brigger, P., Casas, J.R., Pardas, M.: Morphological operators for image and video compression. IEEE Trans. on Image Processing 5(6), 881–898 (1996)
11. `http://www.xilinx.com/support/documentation/v4sx35-video-sk.htm`
12. `http://www.xilinx.com/ipcenter/processor_central/microblaze`
13. de Haan, G.: Motion estimation. In: Video Processing, pp. 221–269. University Press Eindhoven (2004)

Glossary

NTSC National Television System(s) Committee, is an analog television system.

CRT Cathode Ray Tube, is a vacuum tube containing an electron gun (a source of electrons) and a fluorescent screen, with internal or external means to accelerate and deflect the electron beam, used to create images in the form of light emitted from the fluorescent screen.

PAL Phase Alternating Line, is an analogue television encoding system used in broadcast television systems.

SECAM Sequential Color with Memory, is an analog color television system first used in France.

SDTV Standard-definition television, traditional television systems.

HDTV High-Definition Television, is a digital television broadcasting system with high resolution.

CIF Common Intermediate Format, defines a video sequence with a resolution of 352288.

QCIF Quarter CIF, means a quarter part of the CIF resolution, that is, 176x144.

VGA Video Graphics Array, refers to the 640480 resolution.

SVGA Super Video Graphics Array, refers to a resolution of 800600 pixels.

XGA Extended Graphics Array, refers to 1024768 pixels display resolution

MPEG Moving Picture Experts Group, was formed by the ISO to set standards for audio and video compression and transmission.

CIE International Commission on Illumination, the international authority on light, illumination, color, and color spaces.

CCIR International Radio Consultative Committee, is a standard published by International Telecommunication Union - Radiocommunications sector for encoding interlaced analogue video signals in digital form.

ITU International Telecommunication Union, is the second-oldest international organization established to standardize and regulate international radio and telecommunications.

CCD Charge-coupled device, an electronic light sensor used in digital cameras.

Non-MC Non-Motion Compensated de-interlacing algorithms.

ELA Edge-based Line Average, one of the first edge-adaptive de-interlacing algorithms that can use 3+3 or 5+5 taps of the adjacent lines.

VT Vertico-Temporal, is a kind of spatio-temporal de-interlacing algorithm.

MC Motion Compensated de-interlacing algorithms.

SAD Summed Absolute Difference.

3-D RS 3-D Recursive block matcher.

MC TBP Motion-Compensated Temporal Backward Projection de-interlacing algorithm.

MC TR Motion-Compensated Time Recursive.

MC AR Motion-Compensated first-order Adaptive Recursive.

GST Generalized Sampling Theorem de-interlacing algorithm.

GST-2D Generalized Sampling Theorem Bi-dimensional de-interlacing algorithm.

LCD Liquid Crystal Display is a thin, flat panel used for electronically displaying information such as text, images, and moving pictures.

DVD Digital Video Disc player, is a device that plays discs produced under both the DVD-Video and DVD-Audio technical standards.

HD-DVD High Definiton Digital Video Disc player, is a device that plays discs produced in HD-DVD technical standard.

AV receiver Audio-Video receiver is a consumer electronics component to amplify sound from a multitude of possible audio sources as well as route video signals to your TV from various sources.

CEA Consumer Electronics Association, is the trade organization for the consumer electronics industry in the United States.

ASICs Application-Specific Integrated Circuits, is an integrated circuit (IC) customized for a particular use.

DSP Digital Signal Processor, is a specialized microprocessor with optimized architecture for fast operational needs of Digital Signal Processing.

FPGA Field-Programmable Gate Array, is an integrated circuit designed to be configured by the designer after manufacturing.

CAD Computer-Aided Design tools, are used by engineers and designers to design systems.

IP core Intellectual Property core, is a reusable unit of logic, cell, or chip layout design.

FM or HM Fuzzy Mean or Height Method, defuzzifcation method.

MSE Mean Square Error.

PSNR Peak Signal-to-Noise Ratio.

POs Primitive Operations.

Index

MIX
Papier aus verantwortungsvollen Quellen
Paper from responsible sources
FSC® C105338

If you have any concerns about our products,
you can contact us on
ProductSafety@springernature.com

In case Publisher is established outside the EU,
the EU authorized representative is:
Springer Nature Customer Service Center GmbH
Europaplatz 3, 69115 Heidelberg, Germany

Printed by Libri Plureos GmbH
in Hamburg, Germany